# MOLECULAR AND ATOMIC CHEMISTRY

# MOLECULAR AND ATOMIC CHEMISTRY

Michael Gonzalez

Oxford Book Company
Jaipur ▪ New Delhi

*Published By*

Oxford Book Company

**Regd. Off.** "Bony Residency", Gate No. 2
Opp. Tilak Public School, Vishweswariya Nagar,
Gopalpura Road, Jaipur - 302018, India
Ph: 0141 - 2761280, Telefax: 2761381
email - oxfordbookjpr@gmail.com
website - oxfordbookcompany.com

**Delhi Office:** 102, 1st Floor, Satyam House,
4327/3, Ansari Road, Darya Ganj,
New Delhi - 110002, India
Ph: 011-45652440

Distributed by
**KEYA INTERNATIONAL PRESS LLP**
E-mail: keyainternationalpressllp@gmail.com

First published in India, 2024



ISBN : 9789355240958



Typeset by Shivangi Computers, Jaipur

Printed and bound in India by Replika Press Pvt. Ltd.

# Contents

# Preface

One of the most notable recent achievements in the department is the determination of the three-dimensional structure of the T-cell receptor, a key component of the immune response. Understanding its structure and function may enable scientists to enhance the effectiveness of the immune system through the development of new, highly targeted therapeutics. This same research team determined the structure of the erythropoietin receptor bound to a small cyclic peptide that mimics the function of the natural ligand for the receptor. In addition to providing insight into the mechanism of signal transduction by erythropoietin, this work has implications for the rational design of small-molecule drugs that stimulate the development of red blood cells. Other significant accomplishments include the structural determination of the active site of the enzyme that regulates the activity of nitric oxide, the discovery of a novel cortical neuropeptide — cortistatin — that induces sleep, and the solution of the structure of superoxide dismutase, a discovery that has shed light on the mechanism by which mutations cause amyotrophic lateral sclerosis, or Lou Gehrig's disease.

Protein engineering is a major activity of research activity in the department. Improved methods for the generation of catalytic antibodies have been developed, and new antibodies that efficiently catalyze disfavored chemical reactions or facilitate enantiomeric synthesis have been obtained. Directed evolution of proteins is being investigated as a means to produce novel therapeutic agents and to provide deeper understanding of the principles of molecular recognition and catalysis. Additionally, antibodies that bind cocaine have been generated and their potential use in the treatment of cocaine addiction is being explored. In the area of metalloenzyme engineering, scientists have introduced new activities into enzymes by manipulating the protein structure or by introducing new metal-binding centers. Directed evolution methods are being used to generate novel RNA and DNA enzymes. The department continues its investigation into the theoretical and experimental understanding of protein folding. In addition, progress continues unabated in areas of molecular modeling and computational drug design. On the experimental side, novel compounds are being developed that can be designed to target specific DNA sequences and could eventually lead to a new class of therapeutic agents.

Numerous advances have been made in the elucidation of the fundamental molecular processes involved in cell cycle regulation. New insights have been obtained into the mechanism by which mitosis is regulated and the role of various cyclins and kinases in these events. Ongoing research is providing a new understanding of transcriptional regulation during cell growth and the molecular basis for DNA and RNA recognition by zinc-finger proteins. Also, the molecular neurobiology group has identified and characterized several new genes with important functions in the central nervous system, including a receptor protein implicated in mediating circadian rhythms and a new neuropeptide hormone homologous to somastostatin.

The book provide a clear understanding of the principles, applications and limitations of the various techniques involved in analytical chemistry.

**— Michael Gonzalez**

# Chapter 1

## Dynamical System

A dynamical system is a concept in mathematics where a fixed rule describes the time dependence of a point in a geometrical space. Examples include the mathematical models that describe the swinging of a clock pendulum, the flow of water in a pipe, and the number of fish each springtime in a lake.

**Figure 1:** The Lorenz attractor is an example of a non-linear dynamical system. Studying this system helped give rise to Chaos theory.

At any given time a dynamical system has a *state* given by a set of real numbers (a vector) that can be represented by a point in an appropriate *state space* (a geometrical manifold). Small changes in the state of the system create small changes in the numbers. The *evolution rule* of the dynamical system is a fixed rule that describes what future states follow from the current state. The rule is deterministic; in other words, for a given time interval only one future state follows from the current state.

The concept of a dynamical system has its origins in Newtonian mechanics. There, as in other natural sciences and engineering disciplines, the evolution rule of dynamical systems is given implicitly by a relation that gives the state of the system only a short time into the future. (The relation is either a differential equation, difference equation or other time scale.) To determine the state for all future times requires iterating the relation many times—each advancing time a small step. The iteration procedure is referred to as *solving the system* or *integrating the system*. Once the system can be solved, given an initial point it is possible to determine all its future positions, a collection of points known as a *trajectory* or *orbit*.

Before the advent of fast computing machines, solving a dynamical system required sophisticated mathematical techniques and could be accomplished only for a small class of dynamical systems. Numerical methods implemented on electronic computing machines have simplified the task of determining the orbits of a dynamical system.

For simple dynamical systems, knowing the trajectory is often sufficient, but most dynamical systems are too complicated to be understood in terms of individual trajectories. The difficulties arise because:

- The systems studied may only be known approximately—the parameters of the system may not be known precisely or terms may be missing from the equations. The approximations used bring into question the validity or relevance of numerical solutions. To address these questions several notions of stability have been introduced in the study of dynamical systems, such as Lyapunov stability or structural stability. The stability of the dynamical system implies that there is a class of models or initial conditions for which the trajectories would be equivalent. The operation for comparing orbits to establish their equivalence changes with the different notions of stability.
- The type of trajectory may be more important than one particular trajectory. Some trajectories may be periodic, whereas others may wander through many different states of the system. Applications often require enumerating these classes or maintaining the system within one class. Classifying all possible trajectories has led to the qualitative study of dynamical systems, that is, properties that do not change under coordinate changes. Linear dynamical systems and systems that have two numbers describing a state are examples of dynamical systems where the possible classes of orbits are understood.

- The behaviour of trajectories as a function of a parameter may be what is needed for an application. As a parameter is varied, the dynamical systems may have bifurcation points where the qualitative behaviour of the dynamical system changes. For example, it may go from having only periodic motions to apparently erratic behaviour, as in the transition to turbulence of a fluid.
- The trajectories of the system may appear erratic, as if random. In these cases it may be necessary to compute averages using one very long trajectory or many different trajectories. The averages are well defined for ergodic systems and a more detailed understanding has been worked out for hyperbolic systems. Understanding the probabilistic aspects of dynamical systems has helped establish the foundations of statistical mechanics and of chaos.

It was in the work of Poincaré that these dynamical systems themes developed.

## Basic Definitions

A dynamical system is a manifold $M$ called the phase (or state) space endowed with a family of smooth evolution functions $\phi^t$ that for any element of $t \in T$, the time, map a point of the phase space back into the phase space. The notion of smoothness changes with applications and the type of manifold. There are several choices for the set $T$. When $T$ is taken to be the reals, the dynamical system is called a *flow*; and if $T$ is restricted to the non-negative reals, then the dynamical system is a *semi-flow*. When $T$ is taken to be the integers, it is a *cascade* or a *map*; and the restriction to the non-negative integers is a *semi-cascade*.

### *Examples*

The evolution function $\Phi^t$ is often the solution of a *differential equation of motion*

$$\dot{x} = v(x).$$

The equation gives the time derivative, represented by the dot, of a trajectory $x(t)$ on the phase space starting at some point $x_0$. The *vector field* $v(x)$ is a smooth function that at every point of the phase space $M$ provides the velocity vector of the dynamical system at that point. (These vectors are not vectors in the phase space $M$, but in the tangent space $T_xM$ of the point $x$.) Given a smooth $\Phi^t$, an autonomous vector field can be derived from it.

There is no need for higher order derivatives in the equation, nor for time dependence in $v(x)$ because these can be eliminated by considering systems of higher dimensions. Other types of differential equations can be used to define the evolution rule:

$$G(x, \dot{x}) = 0$$

is an example of an equation that arises from the modeling of mechanical systems with complicated constraints.

The differential equations determining the evolution function $\Phi^t$ are often ordinary differential equations: in this case the phase space $M$ is a finite dimensional manifold. Many of the concepts in dynamical systems can be extended to infinite-dimensional manifolds—those that are locally Banach spaces—in which case the differential equations are partial differential equations. In the late 20th century the dynamical system perspective to partial differential equations started gaining popularity.

## Linear Dynamical Systems

Linear dynamical systems can be solved in terms of simple functions and the behaviour of all orbits classified. In a linear system the phase space is the $N$-dimensional Euclidean space, so any point in phase space can be represented by a vector with $N$ numbers. The analysis of linear systems is possible because they satisfy a superposition principle: if $u(t)$ and $w(t)$ satisfy the differential equation for the vector field (but not necessarily the initial condition), then so will $u(t) + w(t)$.

### *Flows*

For a flow, the vector field $\Phi(x)$ is a linear function of the position in the phase space, that is,

$$\phi(x) = Ax + b,$$

with $A$ a matrix, $b$ a vector of numbers and $x$ the position vector. The solution to this system can be found by using the superposition principle (linearity). The case $b \neq 0$ with $A = 0$ is just a straight line in the direction of $b$:

$$\Phi^t(x_1) = x_1 + bt.$$

When $b$ is zero and $A \neq 0$ the origin is an equilibrium (or singular) point of the flow, that is, if $x_0 = 0$, then the orbit remains there. For other initial conditions, the equation of motion is given by the exponential of a matrix: for an initial point $x_0$,

$$\Phi^t(x_0) = e^{tA}x_0.$$

When $b = 0$, the eigenvalues of $A$ determine the structure of the phase space. From the eigenvalues and the eigenvectors of $A$ it is possible to determine if an initial point will converge or diverge to the equilibrium point at the origin.

The distance between two different initial conditions in the case $A \neq 0$ will change exponentially in most cases, either converging exponentially fast towards a point, or diverging exponentially fast. Linear systems display sensitive dependence on initial conditions in the case of divergence. For nonlinear systems this is one of the (necessary but not sufficient) conditions for chaotic behaviour.

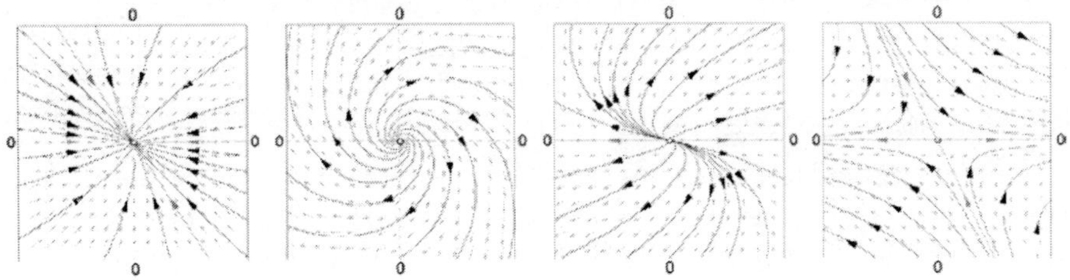

**Figure 2:** Linear vector fields and a few trajectories.

### *Maps*

A discrete-time, affine dynamical system has the form

$$x_{n+1} = Ax_n + b,$$

with $A$ a matrix and $b$ a vector. As in the continuous case, the change of coordinates $x \rightarrow x + (1 - A)^{-1}b$ removes the term $b$ from the equation.

In the new coordinate system, the origin is a fixed point of the map and the solutions are of the linear system $A^n x_0$. The solutions for the map are no longer curves, but points that hop in the phase space. The orbits are organised in curves, or fibres, which are collections of points that map into themselves under the action of the map.

As in the continuous case, the eigenvalues and eigenvectors of $A$ determine the structure of phase space. For example, if $u_1$ is an eigenvector of $A$, with a real eigenvalue smaller than one, then the straight lines given by the points along $a\, u_1$, with $a$ – R, is an invariant curve of the map. Points in this straight line run into the fixed point.

## Local Dynamics

The qualitative properties of dynamical systems do not change under a smooth change of coordinates (this is sometimes taken as a definition of qualitative): a *singular point* of the vector field (a point where $v(x) = 0$) will remain a singular point under smooth transformations; a *periodic orbit* is a loop in phase space and smooth deformations of the phase space cannot alter it being a loop.

It is in the neighbourhood of singular points and periodic orbits that the structure of a phase space of a dynamical system can be well understood. In the qualitative study of dynamical systems, the approach is to show that there is a change of coordinates (usually unspecified, but computable) that makes the dynamical system as simple as possible.

### *Rectification*

A flow in most small patches of the phase space can be made very simple. If $y$ is a point where the vector field $v(y) \neq 0$, then there is a change of coordinates for a region around $y$ where the vector field becomes a series of parallel vectors of the same magnitude. This is known as the rectification theorem.

The rectification theorem says that away from singular points the dynamics of a point in a small patch is a straight line. The patch can sometimes be enlarged by stitching several patches together, and when this works out in the whole phase space $M$ the dynamical system is *integrable*. In most cases the patch cannot be extended to the entire phase space.

There may be singular points in the vector field (where $v(x) = 0$); or the patches may become smaller and smaller as some point is approached. The more subtle reason is a global constraint, where the trajectory starts out in a patch, and after visiting a series of other patches comes back to the original one. If the next time the orbit loops around phase space in a different way, then it is impossible to rectify the vector field in the whole series of patches.

### *Near Periodic Orbits*

In general, in the neighbourhood of a periodic orbit the rectification theorem cannot be used. Poincare developed an approach that transforms the analysis near a periodic orbit to the analysis of a map. Pick a point $x_0$ in the orbit a and consider the points in phase space in that neighbourhood that are perpendicular to $v(x_0)$. These points are a Poincare section $S(\gamma, x_0)$, of the orbit. The flow now defines a map, the Poincare map $F : S \rightarrow S$, for points starting in $S$ and returning to $S$. Not all these points will take the same amount of time to come back, but the times will be close to the time it takes $x_0$.

The intersection of the periodic orbit with the Poincaré section is a fixed point of the Poincare map $F$. By a translation, the point can be assumed to be at $x = 0$. The Taylor series of the map is $F(x) = J \cdot x + O(x^2)$, so a change of coordinates $h$c an only be expected to simplify $F$ to its linear part

$$h^{-1} \circ F \circ h(x) = J \cdot x.$$

This is known as the conjugation equation. Finding conditions for this equation to hold has been one of the major tasks of research in dynamical systems. Poincare first approached it assuming all functions to be analytic and in the process discovered the non-resonant condition. If $\lambda_1, \ldots, \lambda_i$ are the eigenvalues of $J$ they will be resonant if one eigenvalue is an integer linear combination of two or more of the others. As terms of the form $\lambda_i - \Sigma$ (multiples of other eigenvalues) occurs in the denominator of the terms for the function $h$, the non-resonant condition is also known as the small divisor problem.

### Conjugation Results

The results on the existence of a solution to the conjugation equation depend on the eigenvalues of $J$ and the degree of smoothness required from $h$. As $J$ does not need to have any special symmetries, its eigenvalues will typically be complex numbers. When the eigenvalues of $J$ are not in the unit circle, the dynamics near the fixed point $x_0$ of $F$ is called *hyperbolic* and when the eigenvalues are on the unit circle and complex, the dynamics is called *elliptic*.

In the hyperbolic case the Hartman–Grobman theorem gives the conditions for the existence of a continuous function that maps the neighbourhood of the fixed point of the map to the linear map $J \cdot x$. The hyperbolic case is also *structurally stable*. Small changes in the vector field will only produce small changes in the Poincaré map and these small changes will reflect in small changes in the position of the eigenvalues of $J$ in the complex plane, implying that the map is still hyperbolic.

The Kolmogorov–Arnold–Moser (KAM) theorem gives the behaviour near an elliptic point.

### Bifurcation Theory

When the evolution map $\Phi^t$ (or the vector field it is derived from) depends on a parameter $\mu$, the structure of the phase space will also depend on this parameter. Small changes may produce no qualitative changes in the phase space until a special value $\mu_0$ is reached. At this point the phase space changes qualitatively and the dynamical system is said to have gone through a bifurcation.

Bifurcation theory considers a structure in phase space (typically a fixed point, a periodic orbit, or an invariant torus) and studies its behaviour as a function of the parameter $\mu$. At the bifurcation point the structure may change its stability, split into new structures, or merge with other structures. By using Taylor series approximations of the maps and an understanding of the differences that may be eliminated by a change of coordinates, it is possible to catalogue the bifurcations of dynamical systems.

The bifurcations of a hyperbolic fixed point $x_0$ of a system family $F_\mu$ can be characterised by the eigenvalues of the first derivative of the system $DF_\mu(x_0)$ computed at the bifurcation point. For a map, the bifurcation will occur when there are eigenvalues of $DF_\mu$ on the unit circle. For a flow, it will occur when there are eigenvalues on the imaginary axis.

Some bifurcations can lead to very complicated structures in phase space. For example, the Ruelle–Takens scenario describes how a periodic orbit bifurcates into a torus and the torus into a strange attractor. In another example, Feigenbaum period-doubling describes how a stable periodic orbit goes through a series of period-doubling bifurcations.

## Ergodic Systems

In many dynamical systems it is possible to choose the coordinates of the system so that the volume (really a $\upsilon$-dimensional volume) in phase space is invariant. This happens for mechanical systems derived from Newton's laws as long as the coordinates are the position and the momentum and the volume is measured in units of (position) × (momentum). The flow takes points of a subset $A$ into the points $\Phi^t(A)$ and invariance of the phase space means that

$$\mathrm{vol}(A) = \mathrm{vol}(\Phi^t(A)).$$

In the Hamiltonian formalism, given a coordinate it is possible to derive the appropriate (generalised) momentum such that the associated volume is preserved by the flow. The volume is said to be computed by the Liouville measure.

In a Hamiltonian system not all possible configurations of position and momentum can be reached from an initial condition. Because of energy conservation, only the states with the same energy as the initial condition are accessible. The states with the same energy form an energy shell $\Omega$, a sub-manifold of the phase space. The volume of the energy shell, computed using the Liouville measure, is preserved under evolution.

For systems where the volume is preserved by the flow, Poincaré discovered the recurrence theorem: Assume the phase space has a finite Liouville volume and let $F$ be a phase space volume-preserving map and $A$, a subset of the phase space. Then almost every point of $A$ returns to $A$ infinitely often. The Poincaré recurrence theorem was used by Zermelo to object to Boltzmann's derivation of the increase in entropy in a dynamical system of colliding atoms.

One of the questions raised by Boltzmann's work was the possible equality between time averages and space averages, what he called the ergodic hypothesis. The hypothesis states that the length of time a typical trajectory spends in a region $A$ is vol($A$)/vol($\Omega$).

The ergodic hypothesis turned out not to be the essential property needed for the development of statistical mechanics and a series of other ergodic-like properties were introduced to capture the relevant aspects of physical systems. Koopman approached the study of ergodic systems by the use of functional analysis. An observable $a$ is a function that to each point of the phase space associates a number (say instantaneous pressure, or average height). The value of an observable can be computed at another time by using the evolution function $\phi^t$. This introduces an operator $U^t$, the transfer operator,

$$(U^t a)(x) = a(\Phi^{-t}(x)).$$

By studying the spectral properties of the linear operator $U$ it becomes possible to classify the ergodic properties of $\Phi^t$. In using the Koopman approach of considering the action of the flow on an observable function, the finite-dimensional nonlinear problem involving $\Phi^t$ gets mapped into an infinite-dimensional linear problem involving $U$.

The Liouville measure restricted to the energy surface $\Omega$ is the basis for the averages computed in equilibrium statistical mechanics. An average in time along a trajectory is equivalent to an average in space computed with the Boltzmann factor $\exp(-\beta H)$. This idea has been generalised by Sinai, Bowen, and Ruelle (SRB) to a larger class of dynamical systems that includes dissipative systems. SRB measures replace the Boltzmann factor and they are defined on attractors of chaotic systems.

### Nonlinear Dynamical Systems and Chaos

Simple nonlinear dynamical systems and even piecewise linear systems can exhibit a completely unpredictable behaviour, which might seem to be random, despite the fact that they are fundamentally deterministic. This seemingly unpredictable behaviour has been called *chaos*. Hyperbolic systems are precisely defined dynamical systems that exhibit the properties ascribed to chaotic systems. In hyperbolic systems the tangent space perpendicular to a trajectory can be well separated into two parts: one with the points that converge towards the orbit (the *stable manifold*) and another of the points that diverge from the orbit (the *unstable manifold*).

This branch of mathematics deals with the long-term qualitative behaviour of dynamical systems. Here, the focus is not on finding precise solutions to the equations defining the dynamical system (which is often hopeless), but rather to answer questions like "Will the system settle down to a steady state in the long term, and if so, what are the possible attractors?" or "Does the long-term behaviour of the system depend on its initial condition?"

Chaotic behaviour of complex systems is not the issue. Meteorology has been known for years to involve complex—even chaotic—behaviour. Chaos theory has been so surprising because chaos can be found within almost trivial systems. The logistic map is only a second-degree polynomial; the horseshoe map is piecewise linear.

### Geometrical Definition

A dynamical system is the tuple $\langle \mathcal{M}, f, \mathcal{T} \rangle$, with $\mathcal{M}$ a manifold (locally a Banach space or Euclidean space), τ the domain for time (non-negative reals, the integers, ...) and *f* an evolution rule $t \to f^t$ (with $t \in \mathcal{T}$) such that $f^t$ is ad iffeomorphism of the manifold to itself. So, f is a mapping of the time-domain τ into the space of diffeomorphisms of the manifold to itself. In other terms, *f*(*t*) is a diffeomorphism, for every time *t* in the domain τ.

### Measure theoretical definition

A dynamical system may be defined formally, as a measure-preserving transformation of a sigma-algebra, the quadruplet $(X, \Sigma, \mu, \tau)$.

Here, $X$ is a set, and $\Sigma$ is a sigma-algebra on $X$, so that the pair $(X,\Sigma)$ is a measurable space. $\mu$ is a finitemeasure on the sigma-algebra, so that the triplet $(X,\Sigma,\mu)$ is a probability space.

A map $\tau: X \to X$ is said to be $\Sigma$-measurable if and only if, for every $\sigma \in \Sigma$, one has $\tau^{-1}\sigma \in \Sigma$. A map $\tau$ is said to preserve the measure if and only if, for every $\sigma \in \Sigma$, one has $\mu(\tau^{-1}\sigma) = \mu(\sigma)$. Combining the above, a map $\tau$ is said to be a measure-preserving transformation of $X$, if it is a map from $X$ to itself, it is $\Sigma$-measurable, and is measure-preserving. The quadruple$(X,\Sigma,\mu,\tau)$, for such a $\tau$, is then defined to be a dynamical system.

The map $\tau$ embodies the time evolution of the dynamical system. Thus, for discrete dynamical systems the iterates $\tau^n = \tau \circ \tau \circ \ldots \circ \tau$ for integer n are studied. For continuous dynamical systems, the map $\tau$ is understood to be a finite time evolution map and the construction is more complicated.

### Multidimensional Generalisation

Dynamical systems are defined over a single independent variable, usually thought of as time. A more general class of systems are defined over multiple independent variables and are therefore called multidimensional systems. Such systems are useful for modeling, for example, image processing.

## Chemical Thermodynamics

Chemical thermodynamics is the study of the interrelation of heat and work with chemical reactions or with physical changes of state within the confines of the laws of thermodynamics. Chemical thermodynamics involves not only laboratory measurements of various thermodynamic properties, but also the application of mathematical methods to the study of chemical questions and the *spontaneity* of processes.

The structure of chemical thermodynamics is based on the first two laws of thermodynamics. Starting from the first and second laws of thermodynamics, four equations called the “fundamental equations of Gibbs” can be derived. From these four, a multitude of equations, relating the thermodynamic properties of the thermodynamic system can be derived using relatively simple mathematics. This outlines the mathematical framework of chemical thermodynamics.

### History

In 1865, the German physicist Rudolf Clausius, in his *Mechanical Theory of Heat*, suggested that the principles of thermochemistry, e.g. such as the heat evolved in combustion reactions, could be applied to the principles of thermodynamics. Building on the work of Clausius, between the years 1873-76 the American mathematical physicist Willard Gibbs published a series of three papers, the most famous one being the paper *On the Equilibrium of Heterogeneous Substances*. In these papers, Gibbs showed how the first two laws of thermodynamics could be measured graphically and mathematically to determine both the thermodynamic equilibrium of chemical reactions as well as their tendencies to occur or proceed. Gibbs' collection of papers provided the first unified body of thermodynamic theorems from the principles developed by others, such as Clausius and Sadi Carnot.

During the early 20th century, two major publications successfully applied the principles developed by Gibbs to chemical processes, and thus established the foundation of the science of chemical thermodynamics. The first was the 1923 textbook *Thermodynamics and the Free Energy of Chemical Substances* by Gilbert N. Lewis and Merle Randall.

This book was responsible for supplanting the chemical affinity for the term free energy in the English-speaking world. The second was the 1933 book *Modern Thermodynamics by the methods of Willard Gibbs* written by E. A. Guggenheim. In this manner, Lewis, Randall, and Guggenheim are considered as the founders of modern chemical thermodynamics because of the major contribution of these two books in unifying the application of thermodynamics to chemistry.

### Overview

The primary objective of chemical thermodynamics is the establishment of a criterion for the determination of the feasibility or spontaneity of a given transformation. In this manner, chemical thermodynamics is typically used to predict the energy exchanges that occur in the following processes:

1. Chemical reactions
2. Phase changes
3. The formation of solutions.

The following state functions are of primary concern in chemical thermodynamics:

- Internal energy (*U*)
- Enthalpy (*H*).
- Entropy (*S*)
- Gibbs free energy (*G*)

Most identities in chemical thermodynamics arise from application of the first and second laws of thermodynamics, particularly the law of conservation of energy, to these state functions.

The 3 laws of thermodynamics:

1. The energy of the universe is constant.
2. In any spontaneous process, there is always an increase in entropy of the universe
3. The entropy of a perfect crystal at 0 kelvins is zero

## *Chemical Energy*

Chemical energy is the potential of a chemical substance to undergo a transformation through a chemical reaction or to transform other chemical substances. Breaking or making of chemical bonds involves energy, which may be either absorbed or evolved from a chemical system.

Energy that can be released (or absorbed) because of a reaction between a set of chemical substances is equal to the difference between the energy content of the products and the reactants. This change in energy is called the change in internal energy of a chemical reaction. Where $\Delta U^{\circ}_{f\,reactants}$ is the internal energy of formation of the reactant molecules that can be calculated from the bond energies of the various chemical bonds of the molecules under consideration and $\Delta U^{\circ}_{f\,products}$ is the internal energy of formation of the product molecules. The internal energy change of a process is equal to the heat change if it is measured under conditions of constant volume, as in a closed rigid container such as a bomb calorimetre. However, under conditions of constant pressure, as in reactions in vessels open to the atmosphere, the measured heat change is not always equal to the internal energy change, because pressure-volume work also releases or absorbs energy. (The heat change at constant pressure is called the enthalpy change; in this case the enthalpy of formation).

Another useful term is the heat of combustion, which is the energy released due to a combustion reaction and often applied in the study of fuels. Food is similar to hydrocarbon fuel and carbohydrate fuels, and when it is oxidised, its caloric content is similar. In chemical thermodynamics the term used for the chemical potential energy is chemical potential, and for chemical transformation an equation most often used is the Gibbs-Duhem equation.

### Chemical Reactions

In most cases of interest in chemical thermodynamics there are internal degrees of freedom and processes, such as chemical reactions and phase transitions, which always create entropy unless they are at equilibrium, or are maintained at a "running equilibrium" through "quasi-static" changes by being coupled to constraining devices, such as pistons or electrodes, to deliver and receive external work. Even for homogeneous "bulk" materials, the free energy functions depend on the composition, as do all the extensive thermodynamic potentials, including the internal energy. If the quantities $\{ N_i \}$, the number of chemical species, are omitted from the formulae, it is impossible to describe compositional changes.

### Gibbs Function

For a "bulk" (unstructured) system they are the last remaining extensive variables. For an unstructured, homogeneous "bulk" system, there are still various *extensive* compositional variables $\{ N_i \}$ that $G$ depends on, which specify the composition, the amounts of each chemical substance, expressed as the numbers of molecules present or (dividing by Avogadro's number), the numbers of moles

$$G = G(T, P, \{N_i\}).$$

For the case where only $PV$ work is possible

$$dG = -SdT + VdP + \sum_i \mu_i dN_i$$

in which $\mu_i$ is the chemical potential for the $i$-th component in the system

$$\mu_i = \left( \frac{\partial G}{\partial N_i} \right)_{T,P,N_{j \neq i},\text{etc.}}.$$

The expression for d$G$ is especially useful at constant $T$ and $P$, conditions which are easy to achieve experimentally and which approximates the condition in living creatures

$$(dG)_{T,P} = \sum_i \mu_i dN_i.$$

### *Chemical Affinity*

While this formulation is mathematically defensible, it is not particularly transparent since one does not simply add or remove molecules from a system. There is always a *process* involved in changing the composition; e.g., a chemical reaction (or many), or movement of molecules from one phase (liquid) to another (gas or solid). We should find a notation which does not seem to imply that the amounts of the components ( $N_i$ } can be changed independently. All real processes obey conservation of mass, and in addition, conservation of the numbers of atoms of each kind. Whatever molecules are transferred to or from should be considered part of the "system".

Consequently we introduce an explicit variable to represent the degree of advancement of a process, a progress variable $\xi$ for the *extent of reaction*, and to the use of the partial derivative $\partial G/\partial \xi$ (in place of the widely used "$\Delta G$", since the quantity at issue is not a finite change). The result is an understandable expression for the dependence of d$G$ on chemical reactions (or other processes). If there is just one reaction

$$(dG)_{T,P} = \left(\frac{\partial G}{\partial \xi}\right)_{T,P} d\xi.$$

If we introduce the *stoichiometric coefficient* for the *i-th* component in the reaction

$$\nu_i = \partial N_i / \partial \xi$$

which tells how many molecules of $i$ are produced or consumed, we obtain an algebraic expression for the partial derivative

$$\left(\frac{\partial G}{\partial \xi}\right)_{T,P} = \sum_i \mu_i \nu_i = -\mathbb{A}$$

where, we introduce a concise and historical name for this quantity, the "affinity", symbolized by A, as introduced by Theophile de Donder in 1923. The minus sign comes from the fact the affinity was defined to represent the rule that spontaneous changes will ensue only when the change in the Gibbs free energy of the process is negative, meaning that the chemical species have a positive affinity for each other. The differential for $G$ takes on a simple form which displays its dependence on compositional change

$$(dG)_{T,P} = -\mathbb{A}d\xi.$$

If there are a number of chemical reactions going on simultaneously, as is usually the case

$$(dG)_{T,P} = -\sum_k \mathbb{A}_k d\xi_k.$$

a set of reaction coordinates { $\xi_j$ }, avoiding the notion that the amounts of the components ( $N_i$ } can be changed independently. The expressions above are equal to zero at thermodynamic equilibrium, while in the general case for real systems, they are negative because all chemical reactions proceeding at a finite rate produce entropy. This can be made even more explicit by introducing the reaction *rates* $d\xi_j/dt$. For each and every *physically independent process.*

$$\mathbb{A}\dot{\xi} \le 0.$$

This is a remarkable result since the chemical potentials are intensive system variables, depending only on the local molecular milieu. They cannot "know" whether the temperature and pressure (or any other system variables) are going to be held constant over time. It is a purely local criterion and must hold regardless of any such constraints. Of course, it could have been obtained by taking partial derivatives of any of the other fundamental state functions, but nonetheless is a general criterion for ("$T$ times) the entropy production from that spontaneous process; or at least any part of it that is not captured as external work.

We now relax the requirement of a homogeneous "bulk" system by letting the chemical potentials and the affinity apply to any locality in which a chemical reaction (or any other process) is occurring. By accounting for the entropy production due to irreversible processes, the inequality for $dG$ is now replace by an equality

$$dG = -SdT + VdP - \sum_k \mathbb{A}_k d\xi_k + W'$$

or

$$dG_{T,P} = -\sum_k \mathbb{A}_k d\xi_k + W'.$$

Any decrease in the Gibbs function of a system is the upper limit for any isothermal, isobaric work that can be captured in the surroundings, or it may simply be dissipated, appearing as $T$ times a corresponding increase in the entropy of the system and/or its surrounding. Or it may go partly toward doing external work and partly toward creating entropy. The important point is that the *extent of reaction* for a chemical reaction may be coupled to the displacement of some external mechanical or electrical quantity in such a way that one can advance only if the other one also does. The coupling may occasionally be *rigid*, but it is often flexible and variable.

### *Solutions*

In solution chemistry and biochemistry, the Gibbs free energy decrease ("*G*/"ξ, in molar units, denoted cryptically by $\Delta G$) is commonly used as a surrogate for ("*T* times) the entropy produced by spontaneous chemical reactions in situations where there is no work being done; or at least no "useful" work; i.e., other than perhaps some $\pm PdV$. The assertion that all *spontaneous reactions have a negative* $\Delta G$ is merely a restatement of the fundamental thermodynamic relation, giving it the physical dimensions of energy and somewhat obscuring its significance in terms of entropy. When there is no useful work being done, it would be less misleading to use the Legendre transforms of the entropy appropriate for constant *T*, or for constant *T* and *P*, the Massieu functions "*F*/*T* and "*G*/*T* respectively.

### *Non Equilibrium*

Generally the systems treated with the conventional chemical thermodynamics are either at equilibrium or near equilibrium. Ilya Prigogine developed the thermodynamic treatment of open systems that are far from equilibrium. In doing so he has discovered phenomena and structures of completely new and completely unexpected types. His generalised, nonlinear and irreversible thermodynamics has found surprising applications in a wide variety of fields.

The non equilibrium thermodynamics has been applied for explaining how ordered structures e.g. the biological systems, can develop from disorder. Even if Onsager's relations are utilised, the classical principles of equilibrium in thermodynamics still show that linear systems close to equilibrium always develop into states of disorder which are stable to perturbations and cannot explain the occurrence of ordered structures.

Prigogine called these systems dissipative systems, because they are formed and maintained by the dissipative processes which take place because of the exchange of energy between the system and its environment and because they disappear if that exchange ceases.

They may be said to live in symbiosis with their environment. The method which Prigogine used to study the stability of the dissipative structures to perturbations is of very great general interest. It makes it possible to study the most varied problems, such as city traffic problems, the stability of insect communities, the development of ordered biological structures and the growth of cancer cells to mention but a few examples.

### System Constraints

In this regard, it is crucial to understand the role of walls and other *constraints*, and the distinction between *independent* processes and *coupling*. Contrary to the clear implications of many reference sources, the previous analysis is not restricted to homogenous, isotropic bulk systems which can deliver only $PdV$ work to the outside world, but applies even to the most structured systems. There are complex systems with many chemical "reactions" going on at the same time, some of which are really only parts of the same, overall process. An *independent* process is one that *could* proceed even if all others were unaccountably stopped in their tracks. Understanding this is perhaps a "thought experiment" in chemical kinetics, but actual examples exist.

A gas reaction which results in an increase in the number of molecules will lead to an increase in volume at constant external pressure. If it occurs inside a cylinder closed with a piston, the equilibrated reaction can proceed only by doing work against an external force on the piston. The extent variable for the reaction can increase only if the piston moves, and conversely, if the piston is pushed inward, the reaction is driven backwards.

Similarly, a redox reaction might occur in an electrochemical cell with the passage of current in wires connecting the electrodes. The half-cell reactions at the electrodes are constrained if no current is allowed to flow. The current might be dissipated as joule heating, or it might in turn run an electrical device like a motor doing mechanical work. An automobile lead-acid battery can be recharged, driving the chemical reaction backwards. In this case as well, the reaction is not an independent process. Some, perhaps most, of the Gibbs free energy of reaction may be delivered as external work.

The hydrolysis of ATP to ADP and phosphate can drive the force times distance work delivered by living muscles, and synthesis of ATP is in turn driven by a redox chain in mitochondria and chloroplasts, which involves the transport of ionsacross the membranes of these cellular organelles. The coupling of processes here, and in the previous examples, is often not complete. Gas can leak slowly past a piston, just as it can slowly leak out of a rubber balloon. Some reaction may occur in a battery even if no external current is flowing. There is usually a coupling coefficient, which may depend on relative rates, which determines what percentage of the driving free energy is turned into external work, or captured as "chemical work"; a misnomer for the free energy of another chemical process.

## Thermodynamic Databases for Pure Substances

Thermodynamic databases contain information about thermodynamic properties for substances, the most important being enthalpy, entropy, and Gibbs free energy. Numerical values of these thermodynamic properties are collected as tables or are calculated from thermodynamic datafiles. Data is expressed as temperature-dependent values for one mole of substance at the standard pressure of 101.325 kPa (1 atm), or 100 kPa (1 bar). Unfortunately, both of these definitions for the standard condition for pressure are in use.

### *Thermodynamic Data*

Thermodynamic data is usually presented as a table or chart of function values for one mole of a substance (or in the case of the steam tables, one kg). A thermodynamic datafile is a set of equation parameters from which the numerical data values can be calculated. Tables and datafiles are usually presented at a standard pressure of 1 bar or 1 atm, but in the case of steam and other industrially-important gases, pressure may be included as a variable. Function values depend on the state of aggregation of the substance, which must be defined for the value to have any meaning. The state of aggregation for thermodynamic purposes is the *standard state*, sometimes called the *reference state*, and defined by specifying certain conditions. The *normal* standard state is commonly defined as the most stable physical form of the substance at the specified temperature and a pressure of 1 bar or 1 atm. However, since any non-normal condition could be chosen as a standard state, it must be defined in the context of use. A *physical* standard state is one that exists for a time sufficient to allow measurements of its properties. The most common physical standard state is one that is stable thermodynamically (i.e., the normal one). It has no tendency to transform into any other physical state. If a substance can exist but is not thermodynamically stable (for example, a supercooled liquid), it is called a *metastable* state. A *non-physical* standard state is one whose properties are obtained by extrapolation from a physical state (for example, a solid superheated above the normal melting point, or an ideal gas at a condition where the real gas is non-ideal). Metastable liquids and solids are important because some substances can persist and be used in that state indefinitely. Thermodynamic functions that refer to conditions in the normal standard state are designated with a small superscript °. The relationship between certain physical and thermodynamic properties may be described by an equation of state.

### Enthalpy, Heat Content and Heat Capacity

It is very difficult to measure the absolute amount of any thermodynamic quantity involving the internal energy (e.g. enthalpy), since the internal energy of a substance can take many forms, each of which has its own typical temperature at which it begins to become important in thermodynamic reactions. It is therefore the *change* in these functions that is of most interest. The isobaric change in enthalpy $H$ above the common reference temperature of 298.15 K (25 °C) is called the *high temperature heat content*, the *sensible heat*, or the *relative high-temperature enthalpy*, and called henceforth the heat content. Different databases designate this term in different ways; for example $H_T$-$H_{298}$, $H°$-$H°_{298}$, $H°_T$-$H°_{298}$ or $H°$-$H°(T_r)$, where $T_r$ means the reference temperature (usually 298.15 K, but abbreviated in heat content symbols as 298). All of these terms mean the molar heat content for a substance in its normal standard state above a reference temperature of 298.15 K. Data for gases is for the hypothetical ideal gas at the designated standard pressure. The SI unit for enthalpy is J/mol, and is a positive number above the reference temperature. The heat content has been measured and tabulated for virtually all known substances, and is commonly expressed as a polynomial function of temperature. The heat content of an ideal gas is independent of pressure (or volume), but the heat content of real gases varies with pressure, hence the need to define the state for the gas (real or ideal) and the pressure. For some thermodynamic databases such as for steam, the reference temperature is 273.15 K (0 °C).

The *heat capacity* C is the ratio of heat added to the temperature increase. For an incremental isobaric addition of heat:

$$C_P(T) = \left\{ \lim_{\Delta T \to 0} \frac{\Delta H}{\Delta T} \right\} = \left( \frac{\partial H}{\partial T} \right)_p$$

$C_p$ is therefore the slope of a plot of temperature vs. isobaric heat content (or the derivative of a temperature/heat content equation). The SI units for heat capacity are J/(mol·K).

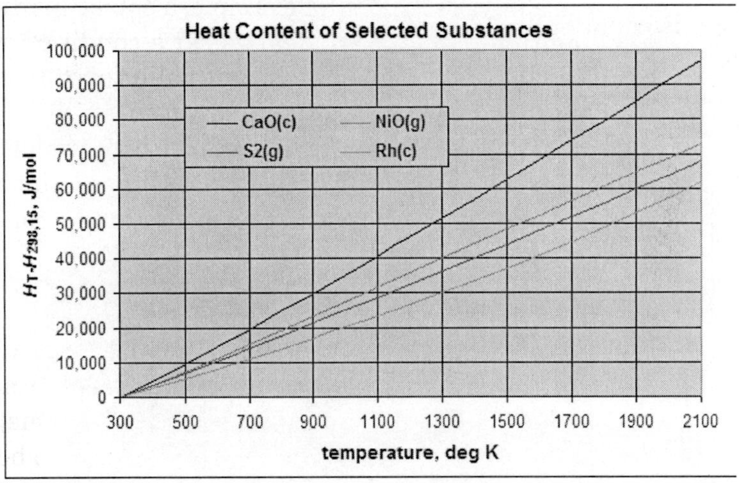


**Figure 3:** Molar heat content of four substances in their designated states above 298.15 K and at 1 atm pressure. CaO(c) and Rh(c) are in their normal standard state of crystalline solid at all temperatures. $S_2$(g) is a non-physical state below about 882 K and NiO(g) is a non-physical state at all temperatures.

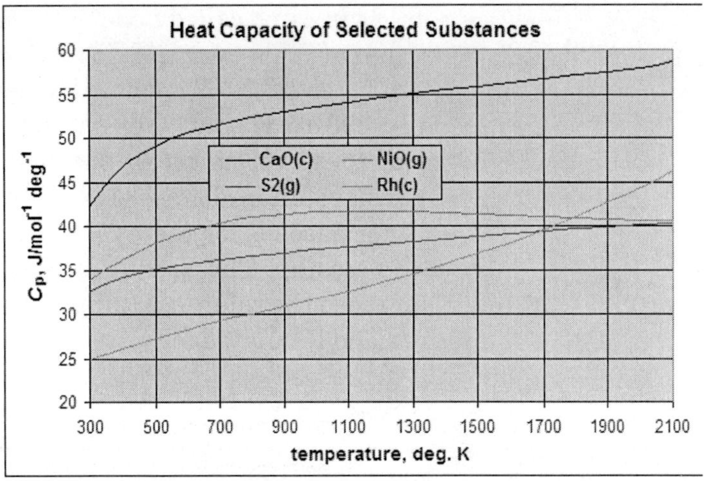


**Figure 4:** Molar heat capacity of four substances in their designated states at 1 atm pressure. CaO(c) and Rh(c) are in their normal standard state of crystalline solid at all temperatures. $S_2$(g) is a non-physical state below about 882 K and NiO(g) is a non-physical state at all temperatures.

### *Enthalpy Change of Phase Transitions*

When heat is added to a condensed-phase substance, its temperature increases until a phase change temperature is reached. With further addition of heat, the temperature remains constant while the phase transition takes place. The amount of substance that transforms is a function of the amount of heat added. After the transition is complete, adding more heat increases the temperature. In other words, the enthalpy of a substance changes isothermally as it undergoes a physical change. The enthalpy change resulting from a phase transition is designated $\Delta H$. There are four types of enthalpy changes resulting from a phase transition. To wit:

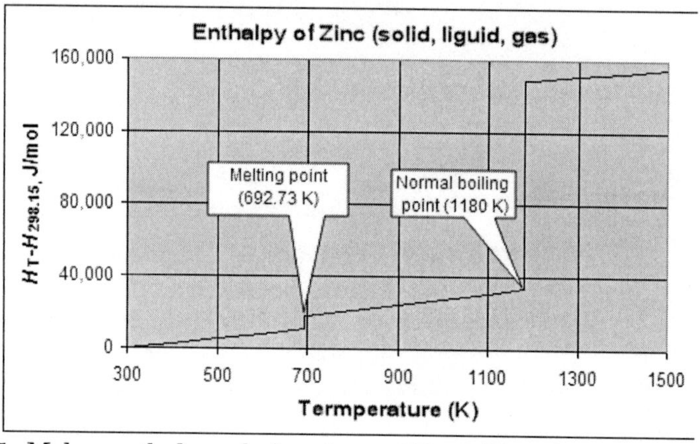


**Figure 5:** Molar enthalpy of zinc above 298.15 K and at 1 atm pressure, showing discontinuities at the melting and boiling points. The $\Delta H°$m of zinc is 7323 J/mol, and the $\Delta H°$v is 115 330 J/mol.

- *Enthalpy of transformation.* This applies to the transformations from one solid phase to another, such as the transformation from a-Fe (bcc ferrite) to a-Fe (fcc austenite). The transformation is designated $\Delta H_{tr}$.
- *Enthalpy of fusion or melting.* This applies to the transition of a solid to a liquid and is designated $\Delta H_{m}$.
- *Enthalpy of vapourisation.* This applies to the transition of a liquid to a vapour and is designated $\Delta H_{v}$.
- *Enthalpy of sublimation.* This applies to the transition of a solid to a vapour and is designated $\Delta H_{s}$.

$C_p$ is infinite at phase transition temperatures because the enthalpy changes isothermally. At the Curie temperature, $C_p$ shows a sharp discontinuity while the enthalpy has a change in slope.

Values of $\Delta H$ are usually given for the transition at the normal standard state temperature for the two states, and if so, are designated with a superscript °. $\Delta H$ for a phase transition is a weak function of temperature. In some texts, the heats of phase transitions are called *latent* heats (for example, *latent heat of fusion*).

## Enthalpy Change for a Chemical Reaction

An enthalpy change occurs during a chemical reaction. For the special case of the formation of a compound from the elements, the change is designated $\Delta H_{form}$ and is a weak function of temperature. Values of $\Delta H_{form}$ are usually given where the elements and compound are in their normal standard states, and as such are designated *standard heats* of formation, as designated by a superscript °. The $\Delta H^{\circ}_{form}$ undergoes discontinuities at a phase transition temperatures of the constituent element(s) and the compound. The enthalpy change for any standard reaction is designated $\Delta H^{\circ}_{rx}$.

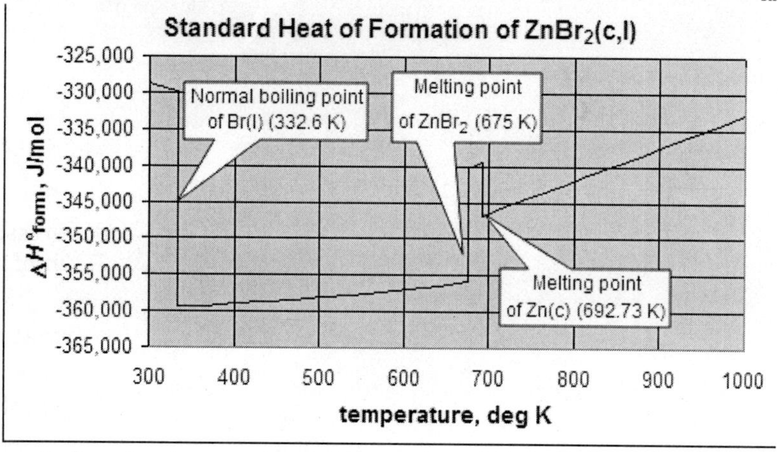


**Figure 6:** Standard molar heat of formation of $ZnBr_2$(c,l) from the elements, showing discontinuities at transition temperatures of the elements and the compound.

## Entropy and Gibbs Energy

The entropy of a system is another thermodynamic quantity that is not easily measured. However, using a combination of theoretical and experimental techniques, entropy can in fact be accurately estimated. At low temperatures, the Debye model leads to the result that the atomic heat capacity $C_v$ for solids should be proportional to $T^3$, and that for perfect crystalline solids it should become zero at absolute zero. Experimentally, the heat capacity is measured at temperature intervals to as low a temperature as possible. Values of $C_p/T$ are plotted against T for the whole range of temperatures where the substance exists in the same physical state. The data are extrapolated from the lowest experimental temperature to 0 K using the Debye model. The third law of thermodynamics states that the entropy of a perfect crystalline substance becomes zero at 0 K. When $S_0$ is zero, the area under the curve from 0 K to any temperature gives the entropy at that temperature. Even though the Debye model contains $C_v$ instead of $C_p$, the difference between the two at temperatures near 0 K is so small as to be negligible.

The absolute value of entropy for a substance in its standard state at the reference temperature of 298.15 K is designated $S°_{298}$. Entropy increases with temperature, and is discontinuous at phase transition temperatures. The change in entropy ($\Delta S°$) at the normal phase transition temperature is equal to the heat of transition divided by the transition temperature. The SI units for entropy are J/(mol·K).

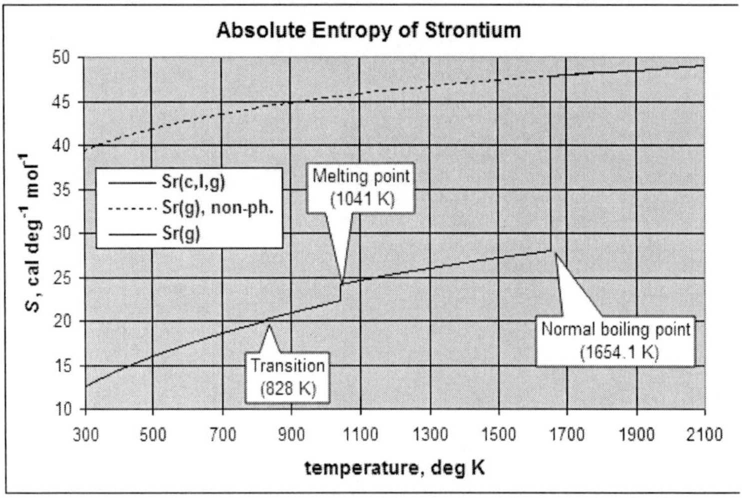


**Figure 7:** Absolute entropy of strontium. The solid line refers to the entropy of strontium in its normal standard state at 1 atm pressure. The dashed line refers to the entropy of strontium vapour in a non-physical state.

The standard entropy change for the formation of a compound from the elements, or for any standard reaction is designated $\Delta S°_{form}$ or $\Delta S°_{rx}$. The entropy change is obtained by summing the absolute entropies of the products minus the sum of the absolute entropies of the reactants. Like enthalpy, the Gibbs energy $G$ has no intrinsic value, so it is the change in $G$ that is of interest. Furthermore, there is no change in $G$ at phase transitions between substances in their standard states. Hence, the main functional application of Gibbs energy from a thermodynamic database is its change in value during the formation of a compound from the standard-state elements, or for any standard chemical reaction ($\Delta G^\circ_{form}$ or $\Delta G^\circ_{rx}$). The SI units of Gibbs energy are the same as for enthalpy (J/mol).

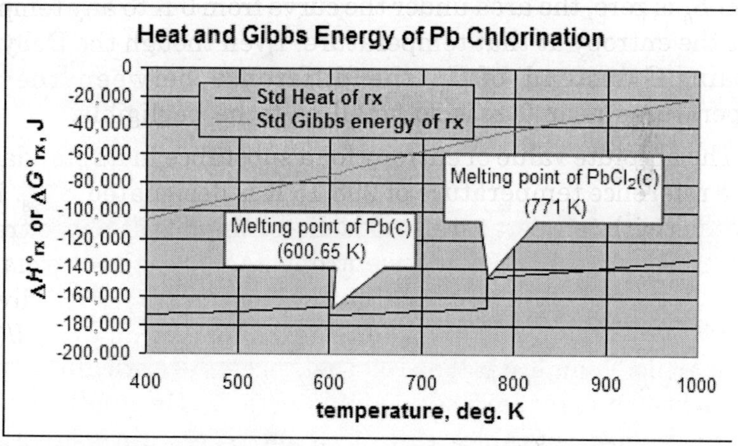


**Figure 8:** Standard heat and Gibbs energy change for the reaction:

$$Pb(c,l) + 2HCl(g) \Rightarrow PbCl_2 + H_2(g)$$

The $\Delta H^\circ_{rx}$ shows discontinuities at the melting points of Pb (600.65 K) and $PbCl_2$ (771 K). $\Delta G^\circ_{rx}$ is not discontinuous at these phase transition temperatures, but does undergo a change in slope, which is almost imperceptible on the chart.

### *Additional Functions*

Compilers of thermochemical databases may contain some additional thermodynamic functions. For example, the absolute enthalpy of a substance $H(T)$ is defined in terms of its formation enthalpy and its heat content as follows:

$$H(T) = \Delta H^\circ_{form,298} + [H_T - H_{298}]$$

For an element, $H(T)$ and $[H_T - H_{298}]$ are identical at all temperatures because $\Delta H^\circ_{form}$ is zero, and of course at 298.15 K, $H(T)$ = 0. For a compound:

$$\Delta H^\circ_{form} = H(T)\text{compound} - \sum\{H(T)\text{elements}\}$$

Similarly, the absolute Gibbs energy $G(T)$ is defined by the absolute enthalpy and entropy of a substance:

$$G(T) = H(T) - T \times S(T)$$

For a compound:

$$\Delta G^{\circ}_{form} = G(T)compound - \sum\{G(T)elements\}$$

Some tables may also contain the Gibbs energy function $(H^{\circ}_{298.15} - G^{\circ}_{T})/T$ which is defined in terms of the entropy and heat content.

$$(H^{\circ}_{298} - G^{\circ}_{T}) / T = S^{\circ}_{T} - (H_{T} - H_{298}) / T$$

The Gibbs energy function has the same units as entropy, but unlike entropy, exhibits no discontinuity at normal phase transition temperatures.

The $\log_{10}$ of the equilibrium constant $K_{eq}$ is often listed, which is calculated from the defining thermodynamic equation.

$$\log_{10}\left(K_{eq}\right) = -\Delta G^{\circ}_{form} / (19.1448T)$$

### Thermodynamic Databases

A thermodynamic database consists of sets of critically evaluated values for the major thermodynamic functions. Originally, data was presented as printed tables at 1 atom and at certain temperatures, usually 100° intervals and at phase transition temperatures. Some compilations included polynomial equations that could be used to reproduce the tabular values. More recently, computerised databases are used which consist of the equation parameters and subroutines to calculate specific values at any temperature and prepare tables for printing. Computerised databases often include subroutines for calculating reaction properties and displaying the data as charts.

Thermodynamic data comes from many types of experiments, such as calorimetry, phase equilibria, spectroscopy, composition measurements of chemical equilibrium mixtures, and emf measurements of reversible reactions. A proper database takes all available information about the elements and compounds in the database, and assures that the presented results are *internally consistent*. Internal consistency requires that all values of the thermodynamic functions are correctly calculated by application of the appropriate thermodynamic equations. For example, values of the Gibbs energy obtained from high-temperature equilibrium emf methods must be identical to those calculated from calorimetric measurements of the enthalpy and entropy values. The database provider must use recognised data analysis procedures to resolve differences between data obtained by different types of experiments. All thermodynamic data is a non-linear function of temperature (and pressure), but there is no universal equation format for expressing the various functions. Here we describe a commonly-used polynomial equation to express the temperature dependence of the heat content. A common six-term equation for the isobaric heat content is:

$$H_T - H_{298} = A(T) + B(T^2) + C(T^{-1}) + D(T^{0.5}) + E(T^3) + F$$

Regardless of the equation format, the heat of formation of a compound at any temperature is $\Delta H°_{form}$ at 298.15 K, plus the sum of the heat content parameters of the products minus the sum of the heat content parameters of the reactants. The $C_p$ equation is obtained by taking the derivative of the heat content equation.

$$C_P = A + 2B(T) - C(T^{-2}) + \frac{1}{2}D(T^{-0.5}) + 3E(T^2)$$

The entropy equation is obtained by integrating the $C_p$/T equation:

$$S_T^\circ = A(\ln T) + 2B(T) + \frac{1}{2}C(T^{-2}) - D(T^{-\frac{1}{2}}) + 1\frac{1}{2}E(T^2) + F'$$

F' is a constant of integration obtained by inserting $S°$ at any temperature $T$. The Gibbs energy of formation of a compound is obtained from the defining equation $\Delta G°_{form} = \Delta H°_{form} - T(\Delta S°_{form})$, and is expressed as

$$\Delta G^\circ_{form} = (\Delta A - \Delta F')T - \Delta A(T \ln T) - \Delta B(T^2) + \frac{1}{2}\Delta C(T^{-1}) + 2\Delta D(T^{\frac{1}{2}})$$

$$-\frac{1}{2}\Delta E(T^3) + \Delta F + \Delta H^\circ_{form298}$$

For most substances, $\Delta G°_{form}$ deviates only slightly from linearity with temperature, so over a short temperature span, the seven-term equation can be replaced by a three-term equation, whose parameter values are obtained by regression of tabular values.

$$\Delta G^\circ_{form} = \alpha + \beta(T \ln T) + \chi$$

Depending on the accuracy of the data and the length of the temperature span, the heat content equation may require more or fewer terms. Over a very long temperature span, two equations may be used instead of one. It is unwise to extrapolate the equations to obtain values outside the range of experimental data used to derive the equation parameters.

## *Thermodynamic Datafiles*

The equation parameters and all other information required to calculate values of the important thermodynamic functions are stored in a thermodynamic datafile.

The values are organised in a format that makes them readable by a thermodynamic calculation program or for use in a spreadsheet. For example, the Excel-based thermodynamic database FREED creates the following type of datafile, here for a standard pressure of 1 atm.

| | | | | | | | | | | |
|---|---|---|---|---|---|---|---|---|---|---|
| | MgCl2 | (c,l,g) | Magnesium Dichloride | | | (K, g-mole, J ) | | | | |
| | Chloromagnesite | | | | | | | | | |
| 1 | 95.2104 | 2.316 | -644294 | 89.5376 | 2000 | | | | | |
| 2 | 3 | | | | | | | | | |
| 3 | 95.0483 | -0.03851 | -1339558 | 0 | 3.15E-05 | | 298.15 | 980 | | |
| 4 | 92.8011 | 0 | 0 | 0 | 0 | | 980 | 1643 | | |
| 5 | 62.1533 | 0 | 0 | 0 | 0 | | 1643 | 2000 | | |
| 6 | 3 | | | | | | | | | |
| 7 | 95.0483 | -0.01925 | 1339558 | 0 | 1.05E-05 | -31397 | | 298.15 | 980 | 39936 |
| 8 | 92.8011 | 0 | 0 | 0 | 0 | 3489 | | 980 | 1643 | 178209 |
| 9 | 62.1533 | 0 | 0 | 0 | 0 | 232053 | | 1643 | 2000 | |
| 10 | 5 | | | | | | | | | |
| 11 | 30.7926 | -0.02421 | 1221780 | 245.164 | 1.03E-05 | -659926 | | 298.15 | 922 | |
| 12 | 27.1692 | -0.02289 | 126828 | 245.164 | 1.03E-05 | -665472 | | 922 | 980 | |
| 13 | 24.922 | -0.00364 | -1212730 | 245.164 | -1.41E-07 | -630585 | | 980 | 1363 | |
| 14 | 29.2775 | 0.000845 | -109828 | 245.164 | -1.41E-07 | -772230 | | 1363 | 1643 | |
| 15 | -1.37026 | 0.000845 | -109828 | 245.164 | -1.41E-07 | -543666 | | 1643 | 2000 | |
| 16 | 5 | | | | | | | | | |
| 17 | -30.7926 | 351.9551 | 0.024215 | 610890 | 490.328 | -5.173E-06 | -659926 | | 298.15 | 922 |
| 18 | -27.1692 | 335.0975 | 0.022893 | 63414 | 490.328 | -5.173E-06 | -665472 | | 922 | 980 |
| 19 | -24.922 | 298.5506 | 0.00364 | -606365 | 490.328 | 7.042E-08 | -630585 | | 980 | 1363 |
| 20 | -29.2775 | 439.7241 | -0.00085 | -54914 | 490.328 | 7.042E-08 | -772230 | | 1363 | 1643 |
| 21 | 1.37026 | 73.68572 | -0.00085 | -54914 | 490.328 | 7.042E-08 | -543666 | | 1643 | 2000 |

**Figure 9:** Thermodynamic datafile for $MgCl_2$(c,l,g) from FREED. Some values have truncated significant figures for display purposes. The explanation for the values is shown below.

- Row 1. Molar mass of species, density at 298.15 K, $\Delta H^{\circ}_{form\ 298.15}$, $S^{\circ}_{298.15}$, and the upper temperature limit for the file.
- Row 2. Number of $C_p$ equations required. Here, three because of three species phases.
- Row 3. Values of the five parameters for the first $C_p$ equation; temperature limit for the equation.
- Row 4. Values of the five parameters for the second $C_p$ equation; temperature limit for the equation.
- Row 5. Values of the five parameters for the third $C_p$ equation; temperature limit for the equation.
- Row 6. Number of $H_T$ - $H_{298}$ equations required.

- Row 7. Values of the six parameters for the first $H_T$ - $H_{298}$ equation; temperature limit for the equation, and $\Delta H^\circ_{trans}$ for the first phase change.
- Row 8. Values of the six parameters for the second $H_T$ - $H_{298}$ equation; temperature limit for the equation, and $\Delta H^\circ_{trans}$ for the second phase change.
- Row 9. Values of the six parameters for the third $H_T$ - $H_{298}$ equation; temperature limit for the equation, and $\Delta H^\circ_{trans}$ for the third phase change.
- Row 10. Number of $\Delta H^\circ_{form}$ equations required. Here five; three for species phases and two because one of the elements has a phase change.
- Row 11. Values of the six parameters for the first $\Delta H^\circ_{form}$ equation; temperature limit for the equation.
- Row 12. Values of the six parameters for the second $\Delta H^\circ_{form}$ equation; temperature limit for the equation.
- Row 13. Values of the six parameters for the third $\Delta H^\circ_{form}$ equation; temperature limit for the equation.
- Row 14. Values of the six parameters for the fourth $\Delta H^\circ_{form}$ equation; temperature limit for the equation.
- Row 15. Values of the six parameters for the fifth $\Delta H^\circ_{form}$ equation; temperature limit for the equation.
- Row 16. Number of $\Delta G^\circ_{form}$ equations required.
- Row 17. Values of the seven parameters for the first $\Delta G^\circ_{form}$ equation; temperature limit for the equation.
- Row 18. Values of the seven parameters for the second $\Delta G^\circ_{form}$ equation; temperature limit for the equation.
- Row 19. Values of the seven parameters for the third $\Delta G^\circ_{form}$ equation; temperature limit for the equation.
- Row 20. Values of the seven parameters for the fourth $\Delta G^\circ_{form}$ equation; temperature limit for the equation.
- Row 21. Values of the seven parameters for the fifth $\Delta G^\circ_{form}$ equation; temperature limit for the equation.

Most computerised databases will create a table of thermodynamic values using the values from the datafile. For $MgCl_2$(c,l,g) at 1 atm pressure.

The FREED table gives additional information in the top rows, such as the mass and amount composition and transition temperatures of the constituent elements. Transition temperatures for the constituent elements have dashes —in the first column in a blank row, such as at 922 K, the melting point of Mg. Transition temperatures for the substance have two blank rows with dashes, and a center row with the defined transition and the enthalpy change, such as the melting point of $MgCl_2$ at 980 K.

## Porphyrin

Porphyrins are a group of organic compounds, many naturally occurring. One of the best-known porphyrins is heme, the pigment in red blood cells; heme is a cofactor of the protein hemoglobin. Porphyrins are heterocyclic macrocycles composed of four modified pyrrole subunits interconnected at their a carbon atoms via methine bridges (=CH-). Porphyrins are aromatic. That is, they obey Hückel's rule for aromaticity, possessing 4n+2 ð electrons (n=4 for the shortest cyclic path) delocalised over the macrocycle. Thus porphyrin macrocycles are highly conjugated systems. As a consequence, they typically have very intense absorption bands in the visible region and may be deeply coloured; the name porphyrin comes from a Greek word for *purple*. The macrocycle has 26 pi electrons in total. The parent porphyrin is porphine, and substituted porphines are called porphyrins.

### *Complexes of Porphyrins and Related Molecules*

Porphyrins are the conjugate acids of ligands that bind metals to form complexes. The metal ion usually has a charge of 2+ or 3+. A schematic equation for these syntheses is shown:

$H_2$porphyrin + $[ML_n]^{2+} \rightarrow$ M(porphyrinate)$L_{n-4}$ + 4 L + 2 $H^+$ where M=metal ion and L=a ligand

A porphyrin without metal in its cavity is a *free base*. Some iron-containing porphyrins are called hemes. Heme-containing proteins, or *hemoproteins*, are found extensively in nature. Hemoglobin and myoglobin are two $O_2$-binding proteins that contain iron porphyrins. Various cytochromes are also hemoproteins.

Several other heterocycles are related to porphyrins. These include corrins, chlorins, bacteriochlorophylls, and corphins. Chlorins (2,3-dihydroporphyrin) are more reduced, contain more hydrogen than porphyrins, and feature a pyrroline subunit. This structure occurs in a chlorophyll molecule. Replacement of two of the four pyrrolic subunits with pyrrolinic subunits results in either a bacteriochlorin (as found in some photosynthetic bacteria) or an isobacteriochlorin, depending on the relative positions of the reduced rings. Some porphyrin derivatives follow Hückel's rule, but most do not.

## Synthesis

***Biosynthesis:*** The "committed step" for porphyrin biosynthesis is the formation of a-aminolevulinic acid (a-ALA, 5-ALA or dALA) by the reaction of the amino acid glycine with succinyl-CoA from the citric acid cycle. Two molecules of dALA combine to give porphobilinogen (PBG), which contains a pyrrole ring.

Four PBGs are then combined through deamination into hydroxymethyl bilane (HMB), which is hydrolysed to form the circular tetrapyrrole uroporphyrinogen III. This molecule undergoes a number of further modifications.

Intermediates are used in different species to form particular substances, but, in humans, the main end-product protoporphyrin IX is combined with iron to form heme. Bile pigments are the breakdown products of heme. The following scheme summarizes the biosynthesis of porphyrins, with references by EC number and the OMIM database. The porphyria associated with the deficiency of each enzyme is also shown:

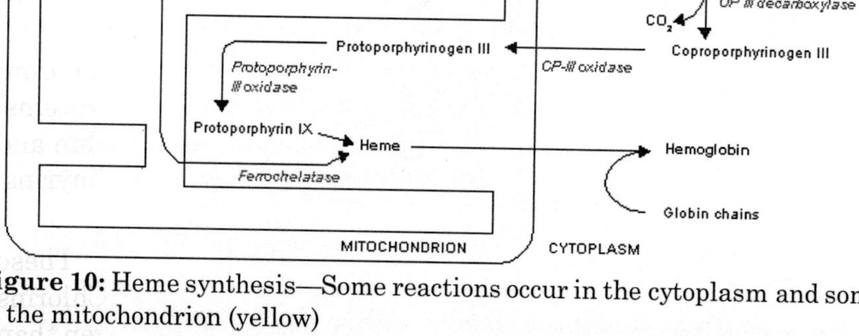


**Figure 10:** Heme synthesis—Some reactions occur in the cytoplasm and some in the mitochondrion (yellow)

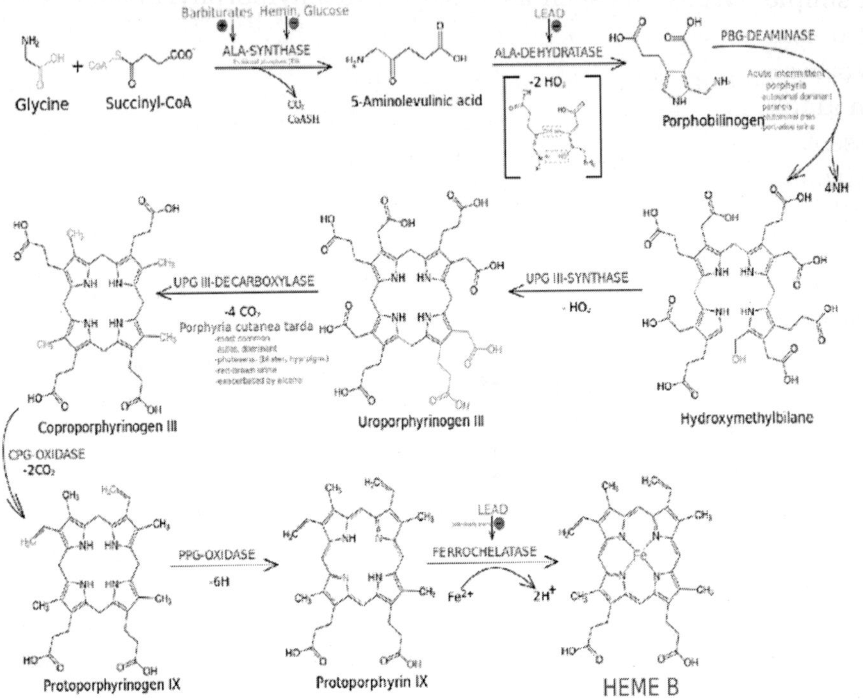


**Figure 11:** Heme B biosynthesis pathway and its modulators. Major enzyme deficiences are also shown here.

### *Laboratory Synthesis*

**Figure 12:** Brilliant crystals of *meso*-tetratolylporphyrin, prepared from 4-methylbenzaldehyde and pyrrole in refluxingpropionic acid

One of the more common syntheses for porphyrins is based on work by Paul Rothemund. His techniques underpin more modern syntheses such as those described by Adler and Longo. The synthesis of simple porphyrins such as *meso*-tetraphenylporphyrin ($H_2TPP$) is also commonly done in university teaching labs. The Rothemund synthesis is an condensation and oxidation starting with pyrrole and an aldehyde. In solution-phase synthesis, acidic conditions are essential; formic acid, acetic acid, and propionic acid are typical reaction solvents, or *p*-toluenesulfonic acid or various Lewis acids can be used with a non-acidic solvent. A large amount of side-product is formed and is removed, usually by recrystallisation or chromatography.

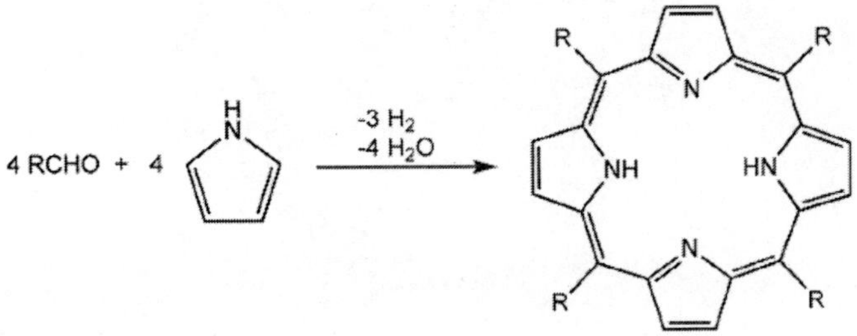


**Figure 13:** Modern green chemistry variants have been developed in which the reaction is performed adsorbed on acidic silica gel with microwave irradiation or at high temperature in the gas phase. In these cases, no additional acid is required.

### Applications

Although natural porphyrin complexes are essential for life, synthetic porphyrins and their complexes have limited utility. Complexes of meso-tetraphenylporphyrin, e.g., the iron(III) chloride complex (TPPFeCl), catalyse a variety of reactions in organic synthesis, but none of them are of practical value. Porphyrin-based compounds are of interest in molecular electronics and supramolecular building blocks. Phthalocyanines, which are structurally related to porphyrins, are used in commerce as dyes and catalysts. Synthetic porphyrin dyes that are incorporated in the design of solar cells are the subject of ongoing research.

In 2008, the corporation Destiny Pharma Ltd reported successful clinical trials of an intra-nasally applied porphyrin XF-73 against methicillin-resistant *Staphylococcus aureus*.

### Supramolecular Chemistry

Porphyrins are often used to construct structures in supramolecular chemistry. These systems take advantage of the Lewis acidity of the metal, typically zinc. An example of a host-guest complex that was constructed from a macrocycle composed of four porphyrins. A guest-free base porphyrin is bound to the center by coordination with its four-pyridine substituents.

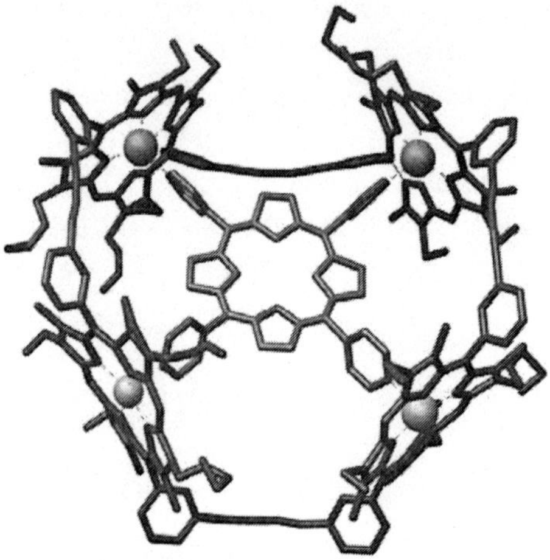

**Figure 14:** An example of a porphyrins involved in host-guest chemistry. Here, a four-porphyrin-zinc complex hosts a porphyrin guest.

## *Organic Geochemistry*

The field of organic geochemistry, the study of the impacts and processes that organisms have had on the Earth, had its origins in the isolation of porphyrins from petroleum. This finding helped establish the biological origins of petroleum. Petroleum is sometimes "fingerprinted" by analysis of trace amounts of nickel and vanadyl porphyrins.

Chlorophyll is a magnesium porphyrin, and heme is an iron porphyrin, but neither porphyrin is present in petroleum. On the other hand, nickel and vanadyl porphyrins could be related to catalytic molecules from bacteria that feed primordial hydrocarbons.

# Chapter 2

## Hydrogen Bond

A hydrogen bond is the attractive interaction of a hydrogen atom with an electronegative atom, such as nitrogen, oxygen or fluorine, that comes from another molecule or chemical group. The hydrogen must be covalently bonded to another electronegative atom to create the bond. These bonds can occur between molecules (*intermolecularly*), or within different parts of a single molecule (*intramolecularly*). The hydrogen bond (5 to 30 kJ/mole) is stronger than a van der Waals interaction, but weaker than covalent or ionic bonds. This type of bond occurs in both inorganic molecules such as water and organic molecules such as DNA.

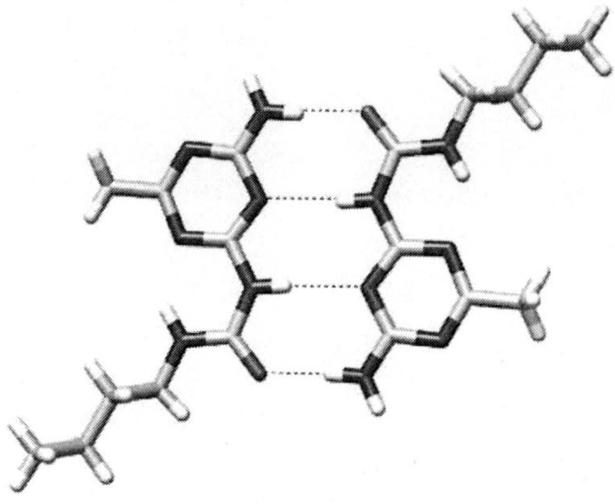

**Figure 1:** An example of intermolecular hydrogen bonding in a self-assembled dimer complex reported by Meijer and coworkers.

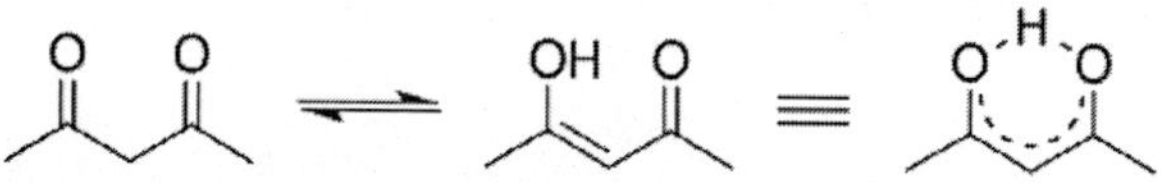


**Figure 2:** Intramolecular hydrogen bonding in acetylacetone helps stabilise the enol tautomer

Intermolecular hydrogen bonding is responsible for the high boiling point of water (100 °C) compared to the other group 16 hydrides that have no hydrogen bonds. Intramolecular hydrogen bonding is partly responsible for the secondary, tertiary, and quaternary structures of proteins and nucleic acids. It also plays an important role in the structure of polymers, both synthetic and natural.

A task group formed by IUPAC has come up with a modern definition of hydrogen bonding in 2011. This new definition can be seen in the IUPAC journal Pure and Applied Chemistry. A detailed technical report provides the rationale behind the new definition.

## Bonding

A hydrogen atom attached to a relatively electronegative atom is a hydrogen bond donor. This electronegative atom is usually fluorine, oxygen, or nitrogen. An electronegative atom such as fluorine, oxygen, or nitrogen is a hydrogen bond acceptor, regardless of whether it is bonded to a hydrogen atom or not. An example of a hydrogen bond donor is ethanol, which has a hydrogen bonded to oxygen; an example of a hydrogen bond acceptor which *does not* have a hydrogen atom bonded to it is the oxygen atom on diethyl ether.

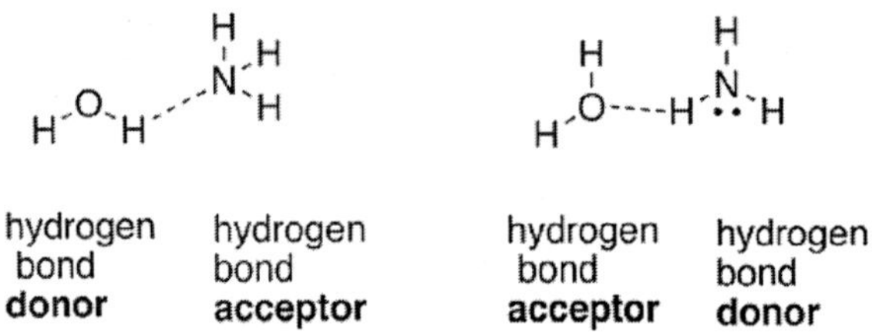


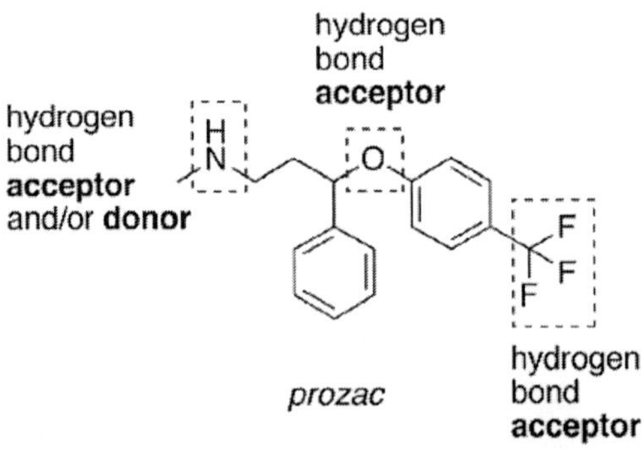


**Figure 3:** Examples of hydrogen bond donating (donors) and hydrogen bond accepting groups (acceptors)

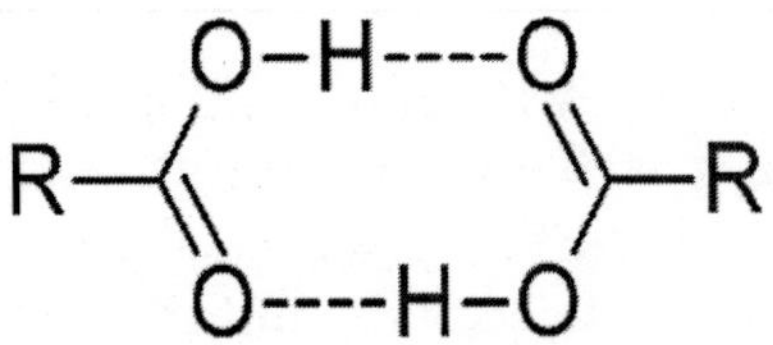


**Figure 4:** Carboxylic acids often form dimers in vapour phase.

A hydrogen attached to carbon can also participate in hydrogen bonding when the carbon atom is bound to electronegative atoms, as is the case in chloroform, $CHCl_3$. The electronegative atom attracts the electron cloud from around the hydrogen nucleus and, by decentralizing the cloud, leaves the atom with a positive partial charge. Because of the small size of hydrogen relative to other atoms and molecules, the resulting charge, though only partial, represents a large charge density. A hydrogen bond results when this strong positive charge density attracts a lone pair of electrons on another heteroatom, which becomes the hydrogen-bond *Acceptor*.

The hydrogen bond is often described as an electrostatic dipole-dipole interaction. However, it also has some features of covalent bonding: it is directional and strong, produces interatomic distances shorter than sum of van der Waals radii, and usually involves a limited number of interaction partners, which can be interpreted as a type of valence. These covalent features are more substantial when acceptors bind hydrogens from more electronegative donors.

The partially covalent nature of a hydrogen bond raises the following questions: "To which molecule or atom does the hydrogen nucleus belong?" and "Which should be labelled 'donor' and which 'acceptor'?" Usually, this is simple to determine on the basis of interatomic distances in the X"H···Y system: X"H distance is typically H"110 pm, whereas H···Y distance is H"160 to 200 pm. Liquids that display hydrogen bonding are called associated liquids.

Hydrogen bonds can vary in strength from very weak (1–2 kJ $mol^{-1}$) to extremely strong (161.5 kJ $mol^{-1}$ in the ion HF"2). Typical enthalpies in vapour include:

- F–H···:F (161.5 kJ/mol or 38.6 kcal/mol)
- O–H···:N (29 kJ/mol or 6.9 kcal/mol)
- O–H···:O (21 kJ/mol or 5.0 kcal/mol)
- N–H···:N (13 kJ/mol or 3.1 kcal/mol)

- N–H···:O (8 kJ/mol or 1.9 kcal/mol)
- HO–H···:OH+3 (18 kJ/mol or 4.3 kcal/mol; data obtained using molecular dynamics as detailed in the reference and should be compared to 7.9 kJ/mol for bulk waters, obtained using the same molecular dynamics.)

The length of hydrogen bonds depends on bond strength, temperature, and pressure. The bond strength itself is dependent on temperature, pressure, bond angle, and environment (usually characterised by local dielectric constant). The typical length of a hydrogen bond in water is 197 pm. The ideal bond angle depends on the nature of the hydrogen bond donor. The following hydrogen bond angles between a hydrofluoric acid donor and various acceptors have been determined experimentally:

| *Acceptor···donor* | *VSEPR symmetry* | *Angle (°)* |
|---|---|---|
| HCN···HF | linear | 180 |
| $H_2CO$···HF | trigonal planar | 110 |
| $H_2O$···HF | pyramidal | 46 |
| $H_2S$···HF | pyramidal | 89 |
| $SO_2$···HF | trigonal | 142 |

### *History*

In the book *The Nature of the Chemical Bond*, Linus Pauling credits T. S. Moore and T. F. Winmill with the first mention of the hydrogen bond, in 1912. Moore and Winmill used the hydrogen bond to account for the fact that trimethylammonium hydroxide is a weaker base than tetramethylammonium hydroxide. The description of hydrogen bonding in its more well-known setting, water, came some years later, in 1920, from Latimer and Rodebush. In that paper, Latimer and Rodebush cite work by a fellow scientist at their laboratory, Maurice Loyal Huggins, saying, "Mr. Huggins of this laboratory in some work as yet unpublished, has used the idea of a hydrogen kernel held between two atoms as a theory in regard to certain organic compounds."

## Hydrogen Bonds in Water

The most ubiquitous, and perhaps simplest, example of a hydrogen bond is found between water molecules. In a discrete water molecule, there are two hydrogen atoms and one oxygen atom. Two molecules of water can form a hydrogen bond between them; the simplest case, when only two molecules are present, is called the water dimer and is often used as a model system. When more molecules are present, as is the case of liquid water, more bonds are possible because the oxygen of one water molecule has two lone pairs of electrons, each of which can form a hydrogen bond with a hydrogen on another water molecule.

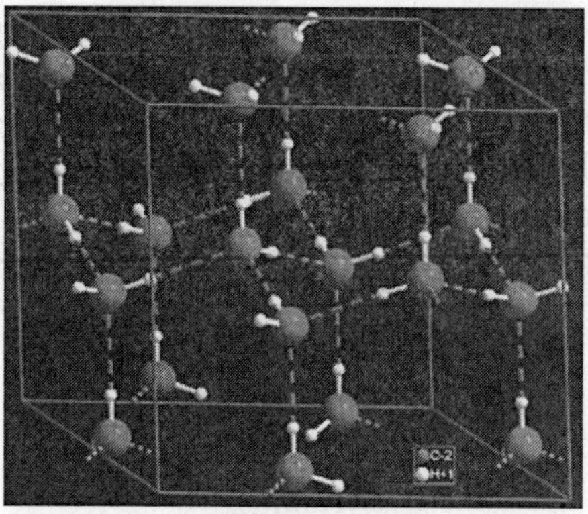

**Figure 5:** Crystal structure of hexagonal ice. Gray dashed lines indicate hydrogen bonds

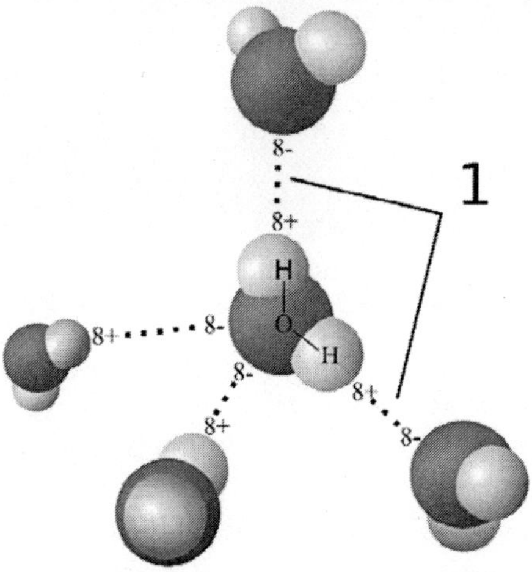


**Figure 6:** Model of hydrogen bonds (1) between molecules of water

This can repeat such that every water molecule is H-bonded with up to four other molecules. Hydrogen bonding strongly affects the crystal structure of ice, helping to create an open hexagonal lattice. The density of ice is less than water at the same temperature; thus, the solid phase of water floats on the liquid, unlike most other substances. Liquid water's high boiling point is due to the high number of hydrogen bonds each molecule can form relative to its low molecular mass. Owing to the difficulty of breaking these bonds, water has a very high boiling point, melting point, and viscosity compared to otherwise similar liquids not conjoined by hydrogen bonds. Water is unique because its oxygen atom has two lone pairs and two hydrogen atoms, meaning that the total number of bonds of a water molecule is up to four. For example, hydrogen fluoride—which has three lone pairs on the F atom but only one H atom—can form only two bonds; (ammonia has the opposite problem: three hydrogen atoms but only one lone pair).

H–F···H–F···H–F

The exact number of hydrogen bonds formed by a molecule of liquid water fluctuates with time and depends on the temperature. From TIP4P liquid water simulations at 25 °C, it was estimated that each water molecule participates in an average of 3.59 hydrogen bonds. At 100 °C, this number decreases to 3.24 due to the increased molecular motion and decreased density, while at 0 °C, the average number of hydrogen bonds increases to 3.69. A more recent study found a much smaller number of hydrogen bonds: 2.357 at 25 °C. The differences may be due to the use of a different method for defining and counting the hydrogen bonds.

Where the bond strengths are more equivalent, one might instead find the atoms of two interacting water molecules partitioned into two polyatomic ions of opposite charge, specifically hydroxide ($OH^-$) and hydronium ($H_3O^+$). (Hydronium ions are also known as 'hydroxonium' ions.)

$H''O^- H_3O^+$

Indeed, in pure water under conditions of standard temperature and pressure, this latter formulation is applicable only rarely; on average about one in every $5.5 \times 10^8$ molecules gives up a proton to another water molecule, in accordance with the value of the dissociation constant for water under such conditions. It is a crucial part of the uniqueness of water.

Because water forms hydrogen bonds with the donors and acceptors on solutes dissolved within it, it inhibits the formation of a hydrogen bond between two molecules of those solutes or the formation of intramolecular hydrogen bonds within those solutes through competition for their donors and acceptors. Consequently, hydrogen bonds between or within solute molecules dissolved in water are almost always unfavourable relative to hydrogen bonds between water and the donors and acceptors for hydrogen bonds on those solutes.

### *Bifurcated and Over-coordinated Hydrogen Bonds in Water*

It can be that a single hydrogen atom participates in two hydrogen bonds, rather than one. This type of bonding is called "bifurcated" (split in two or 'two-forked'). It can exist for instance in complex natural or synthetic organic molecules It was suggested that a bifurcated hydrogen atom is an essential step in water reorientation.

Acceptor type hydrogen bonds (terminating on an oxygen's lone pairs), are more likely to form bifurcation (it is called overcoordinated oxygen, OCO) than donor type, beginning on the same oxygen's hydrogens.

## Hydrogen Bonds in DNA and Proteins

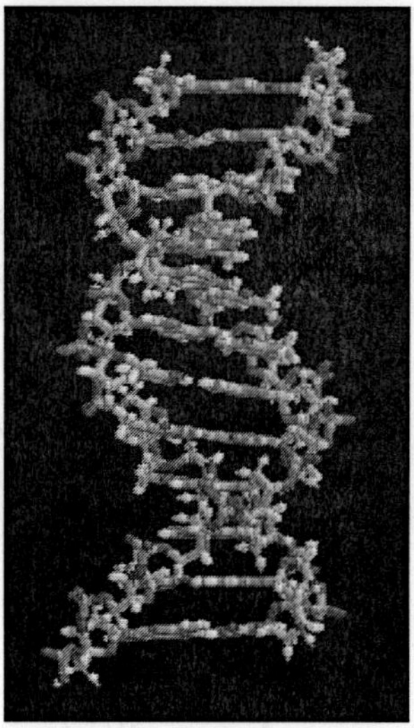

**Figure 7:** The structure of part of a DNA double helix

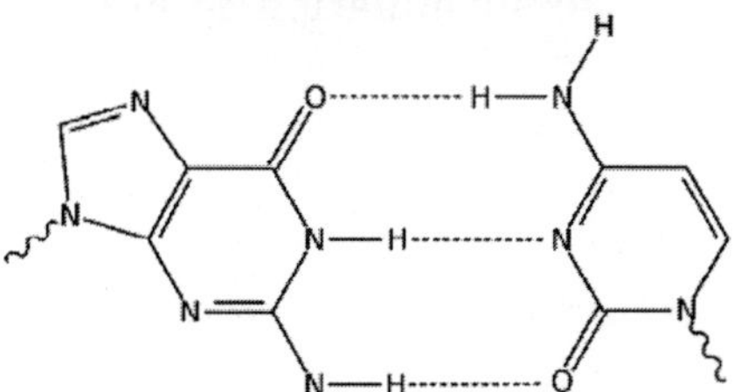


**Figure 8:** Hydrogen bonding between guanine andcytosine, one of two types of base pairs in DNA.

Hydrogen bonding also plays an important role in determining the three-dimensional structures adopted by proteins and nucleic bases. In these macromolecules, bonding between parts of the same macromolecule cause it to fold into a specific shape, which helps determine the molecule's physiological or biochemical role. The double helical structure of DNA, for example, is due largely to hydrogen bonding between the base pairs, which link one complementary strand to the other and enable replication.

In the secondary structure of proteins, hydrogen bonds form between the backbone oxygens and amide hydrogens. When the spacing of the amino acid residues participating in a hydrogen bond occurs regularly between positions $i$ and $i + 4$, an alpha helix is formed. When the spacing is less, between positions $i$ and $i + 3$, then a $3_{10}$ helix is formed. When two strands are joined by hydrogen bonds involving alternating residues on each participating strand, a beta sheet is formed. Hydrogen bonds also play a part in forming the tertiary structure of protein through interaction of R-groups.

### Hydrogen Bonds in Polymers

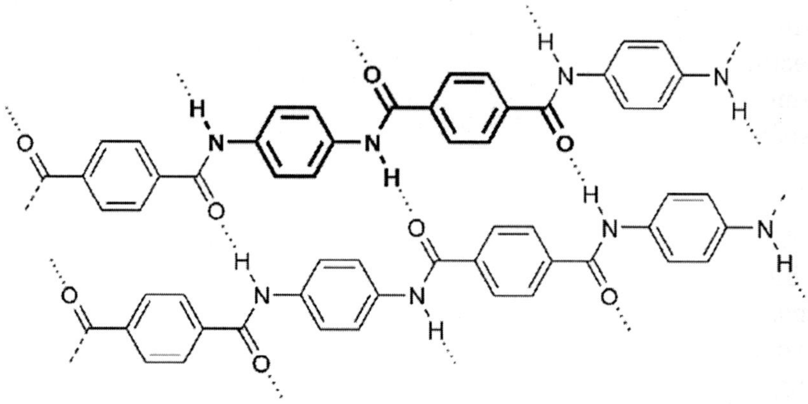

**Figure 9:** Para-aramid structure

Many polymers are strengthened by hydrogen bonds in their main chains. Among the synthetic polymers, the best known example is nylon, where hydrogen bonds occur in the repeat unit and play a major role in crystallisation of the material. The bonds occur between carbonyl and amine groups in the amide repeat unit. They effectively link adjacent chains to create crystals, which help reinforce the material. The effect is greatest in aramid fibre, where hydrogen bonds stabilise the linear chains laterally. The chain axes are aligned along the fibre axis, making the fibres extremely stiff and strong. Hydrogen bonds are also important in the structure of cellulose and derived polymers in its many different forms in nature, such as wood and natural fibres such as cotton and flax.

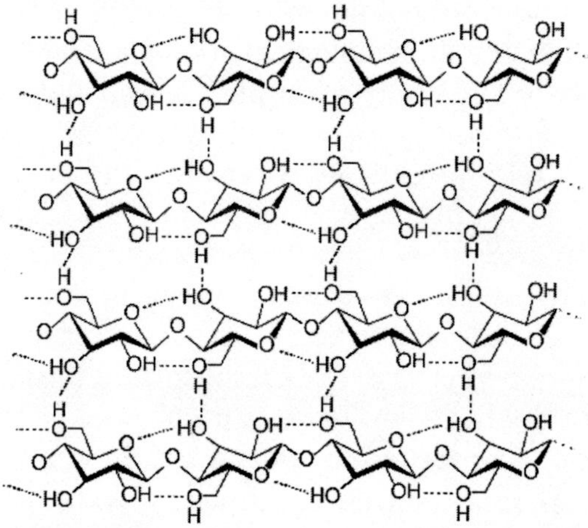

**Figure 10:** A strand of cellulose (conformation $I_{\alpha}$), showing the hydrogen bonds (dashed) within and between cellulose molecules.

The hydrogen bond networks make both natural and synthetic polymers sensitive to humidity levels in the atmosphere because water molecules can diffuse into the surface and disrupt the network. Some polymers are more sensitive than others. Thus nylons are more sensitive than aramids, and nylon 6 more sensitive than nylon-11.

## Symmetric Hydrogen Bond

A symmetric hydrogen bond is a special type of hydrogen bond in which the proton is spaced exactly halfway between two identical atoms. The strength of the bond to each of those atoms is equal. It is an example of a 3-center 4-electron bond. This type of bond is much stronger than "normal" hydrogen bonds. The effective bond order is 0.5, so its strength is comparable to a covalent bond. It is seen in ice at high pressure, and also in the solid phase of many anhydrous acids such as hydrofluoric acid and formic acid at high pressure. It is also seen in the bifluoride ion [F"H"F]$^-$.

Symmetric hydrogen bonds have been observed recently spectroscopically in formic acid at high pressure (>GPa). Each hydrogen atom forms a partial covalent bond with two atoms rather than one. Symmetric hydrogen bonds have been postulated in ice at high pressure

(Ice X). Low-barrier hydrogen bonds form when the distance between two heteroatoms is very small.

## Dihydrogen Bond

The hydrogen bond can be compared with the closely related dihydrogen bond, which is also an intermolecular bonding interaction involving hydrogen atoms. These structures have been known for some time, and well characterised by crystallography; however, an understanding of their relationship to the conventional hydrogen bond, ionic bond, and covalent bond remains unclear. Generally, the hydrogen bond is characterised by a proton acceptor that is a lone pair of electrons in nonmetallic atoms (most notably in the nitrogen, and chalcogen groups). In some cases, these proton acceptors may be pi-bonds or metal complexes. In the dihydrogen bond, however, a metal hydride serves as a proton acceptor; thus forming a hydrogen-hydrogen interaction. Neutron diffraction has shown that the molecular geometry of these complexes is similar to hydrogen bonds, in that the bond length is very adaptable to the metal complex/ hydrogen donor system.

## Advanced Theory of the Hydrogen Bond

In 1999, Isaacs *et al.* proved from interpretations of the anisotropies in the Compton profile of ordinary ice, that the hydrogen bond is partly covalent. Some NMR data on hydrogen bonds in proteins also indicate covalent bonding.

Most generally, the hydrogen bond can be viewed as a metric-dependent electrostatic scalar field between two or more intermolecular bonds. This is slightly different from the intramolecular bound states of, for example, covalent or ionic bonds; however, hydrogen bonding is generally still a bound state phenomenon, since the interaction energy has a net negative sum.

The initial theory of hydrogen bonding proposed by Linus Pauling suggested that the hydrogen bonds had a partial covalent nature. This remained a controversial conclusion until the late 1990s when NMR techniques were employed by F.

Cordier *et al.* to transfer information between hydrogen-bonded nuclei, a feat that would only be possible if the hydrogen bond contained some covalent character. While much experimental data has been recovered for hydrogen bonds in water, for example, that provide good resolution on the scale of intermolecular distances and molecular thermodynamics, the kinetic and dynamical properties of the hydrogen bond in dynamic systems remains unchanged.

## Hydrogen Bonding Phenomena

- Dramatically higher boiling points of $NH_3$, $H_2O$, and HF compared to the heavier analogues $PH_3$, $H_2S$, and HCl.
- Increase in the melting point, boiling point, solubility, and viscosity of many compounds can be explained by the concept of hydrogen bonding.
- Viscosity of anhydrous phosphoric acid and of glycerol
- Dimer formation in carboxylic acids and hexamer formation in hydrogen fluoride, which occur even in the gas phase, resulting in gross deviations from the ideal gas law.
- Pentamer formation of water and alcohols in apolar solvents.
- High water solubility of many compounds such as ammonia is explained by hydrogen bonding with water molecules.
- Negative azeotropy of mixtures of HF and water
- Deliquescence of NaOH is caused in part by reaction of $OH^-$ with moisture to form hydrogen-bonded $H_3O$"

2 species. An analogous process happens between $NaNH_2$ and $NH_3$, and between NaF and HF.

- The fact that ice is less dense than liquid water is due to a crystal structure stabilised by hydrogen bonds.
- The presence of hydrogen bonds can cause an anomaly in the normal succession of states of matter for certain mixtures of chemical compounds as temperature increases or decreases. These compounds can be liquid until a certain temperature, then solid even as the temperature increases, and finally liquid again as the temperature rises over the "anomaly interval"
- Smart rubber utilises hydrogen bonding as its sole means of bonding, so that it can "heal" when torn, because hydrogen bonding can occur on the fly between two surfaces of the same polymer.
- Strength of nylon and cellulose fibres.
- Wool, being a protein fibre is held together by hydrogen bonds, causing wool to recoil when stretched. However, washing at high temperatures can permanently break the hydrogen bonds and a garment may permanently lose its shape.

## Coordination Complex

In chemistry, a coordination complex or metal complex, is an atom or ion (usually metallic), bonded to a surrounding array of molecules or anions, that are in turn known as ligands or complexing agents. Many metal-containing compounds consist of coordination complexes.

### *Nomenclature and Terminology*

Coordination complexes are so pervasive that the structure and reactions are described in many ways, sometimes confusingly. The atom within a ligand that is bonded to the central atom or ion is called the donor atom. A typical complex is bound to several donor atoms, which can be the same or different. Polydentate (multiple bonded) ligands consist of several donor atoms, several of which are bound to the central atom or ion. These complexes are called chelate complexes, the formation of such complexes is called chelation, complexation, and coordination.

The central atom or ion, together with all ligands comprise the coordination sphere. The central atoms or ion and the donor atoms comprise the first coordination sphere.

Coordination refers to the "coordinate covalent bonds" (dipolar bonds) between the ligands and the central atom. Originally, a complex implied a reversible association of molecules, atoms, or ionsthrough such weak chemical bonds. As applied to coordination chemistry, this meaning has evolved. Some metal complexes are formed virtually irreversibly and many are bound together by bonds that are quite strong.

### *History*

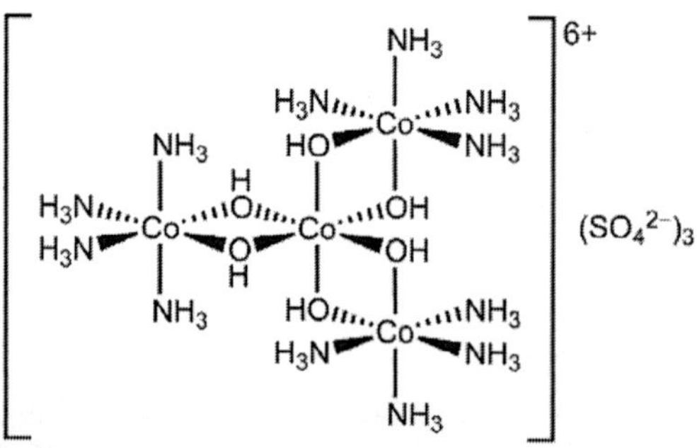


**Figure 11:** Structure of hexol

Coordination complexes were known - although not understood in any sense - since the beginning of chemistry, e.g. Prussian blue and copper vitriol. The key breakthrough occurred when Alfred Werner proposed in 1893 that Co(III) bears six ligands in an octahedral geometry. His theory allows one to understand the difference between coordinated and ionic in a compound, for example chloride in the cobalt ammine chlorides and to explain many of the previously inexplicable isomers.

In 1914, Werner resolved the first coordination complex, called hexol, into optical isomers, overthrowing the theory that chirality was necessarily associated with carbon compounds.

### *Structures*

The ions or molecules surrounding the central atom are called ligands. Ligands are generally bound to the central atom by a coordinate covalent bond (donating electrons from a lone electron pair into an empty metal orbital), and are said to be coordinated to the atom. There are also organic ligands such as alkenes whose pi bonds can coordinate to empty metal orbitals. An example is ethene in the complex known as Zeise's salt, $K^+[PtCl_3(C_2H_4)]^-$.

### *Geometry*

In coordination chemistry, a structure is first described by its coordination number, the number of ligands attached to the metal (more specifically, the number of donor atoms). Usually one can count the ligands attached, but sometimes even the counting can become ambiguous. Coordination numbers are normally between two and nine, but large numbers of ligands are not uncommon for the lanthanides and actinides. The number of bonds depends on the size, charge, and electron configuration of the metal ion and the ligands. Metal ions may have more than one coordination number.

Typically the chemistry of complexes is dominated by interactions between s and p molecular orbitals of the ligands and the d orbitals of the metal ions. The s, p, and d orbitals of the metal can accommodate 18 electrons. The maximum coordination number for a certain metal is thus related to the electronic configuration of the metal ion (to be more specific, the number of empty orbitals) and to the ratio of the size of the ligands and the metal ion. Large metals and small ligands lead to high coordination numbers, e.g. $[Mo(CN)_8]$. Small metals with large ligands lead to low coordination numbers, e.g. $Pt[P(CMe_3)]_2$. Due to their large size, lanthanides, actinides, and early transition metals tend to have high coordination numbers.

Different ligand structural arrangements result from the coordination number. Most structures follow the points-on-a-sphere pattern (or, as if the central atom were in the middle of a polyhedron where the corners of that shape are the locations of the ligands), where orbital overlap (between ligand and metal orbitals) and ligand-ligand repulsions tend to lead to certain regular geometries. The most observed geometries are listed below, but there are many cases that deviate from a regular geometry, e.g. due to the use of ligands of different types (which results in irregular bond lengths; the coordination atoms do not follow a points-on-a-sphere pattern), due to the size of ligands, or due to electronic effects:

- Linear for two-coordination
- Trigonal planar for three-coordination
- Tetrahedral or square planar for four-coordination
- Trigonal bipyramidal or square pyramidal for five-coordination
- Octahedral (orthogonal) or trigonal prismatic for six-coordination
- Pentagonal bipyramidal for seven-coordination
- Square antiprismatic for eight-coordination
- Tri-capped trigonal prismatic (Triaugmented triangular prism) for nine coordination.

Some exceptions and provisions should be noted:

- The idealised descriptions of 5-, 7-, 8-, and 9- coordination are often indistinct geometrically from alternative structures with slightly different L-M-L (ligand-metal-ligand) angles. The classic example of this is the difference between square pyramidal and trigonal bipyramidal structures.
- Due to special electronic effects such as (second-order) Jahn-Teller stabilisation, certain geometries are stabilised relative to the other possibilities, e.g. for some compounds the trigonal prismatic geometry is stabilised relative to octahedral structures for six-coordination.

### *Isomerism*

The arrangement of the ligands is fixed for a given complex, but in some cases it is mutable by a reaction that forms another stable isomer.

There exist many kinds of isomerism in coordination complexes, just as in many other compounds.

### *Stereoisomerism*

Stereoisomerism occurs with the same bonds in different orientations relative to one another. Stereoisomerism can be further classified into:

### *Cis-trans Isomerism and Facial-meridional Isomerism*

Cis-trans isomerism occurs in octahedral and square planar complexes (but not tetrahedral). When two ligands are mutually adjacent they are said to be cis, when opposite each other, trans. When three identical ligands occupy one face of an octahedron, the isomer is said to be facial, or fac. In a *fac* isomer, any two identical ligands are adjacent or *cis* to each other. If these three ligands and the metal ion are in one plane, the isomer is said to be meridional, or mer. A *mer*isomer can be considered as a combination of a *trans* and a *cis*, since it contains both trans and cis pairs of identical ligands.

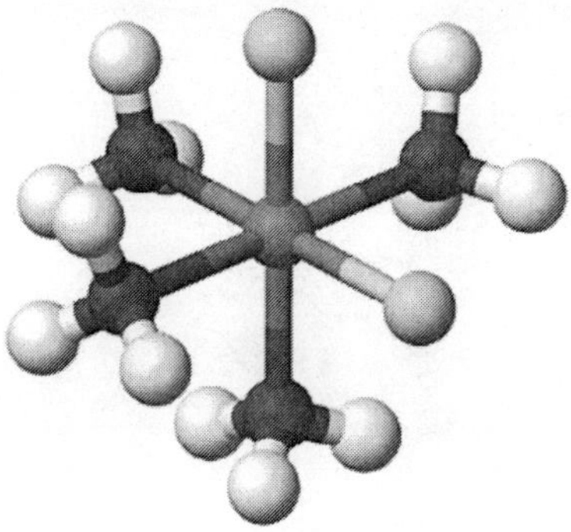

**Figure 12:** *cis*-$[CoCl_2(NH_3)_4]^+$

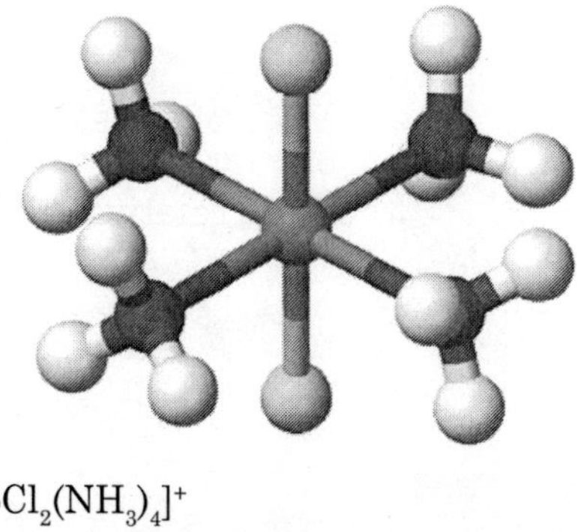

**Figure 13:** *trans*-$[CoCl_2(NH_3)_4]^+$

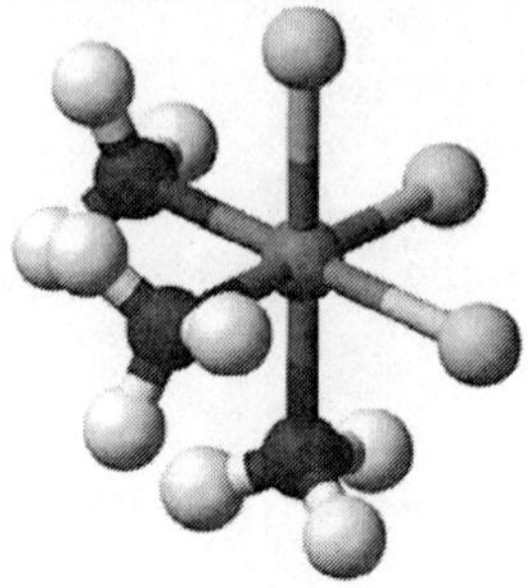

**Figure 14:** *fac*-$[CoCl_3(NH_3)_3]$

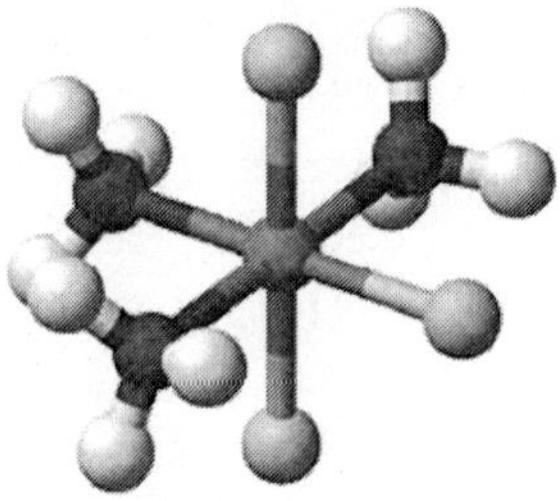

**Figure 15:** *mer*-$[CoCl_3(NH_3)_3]$

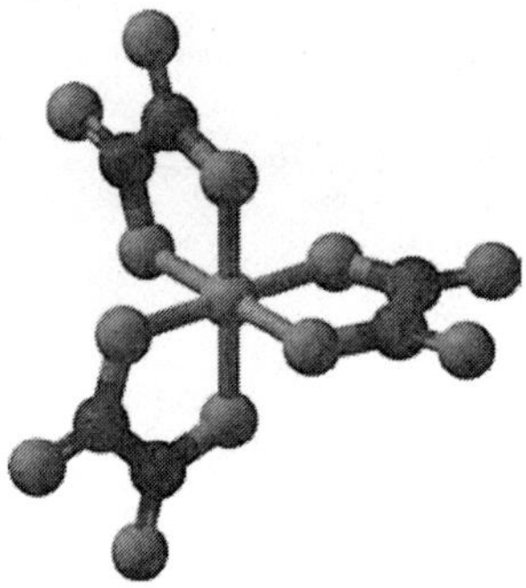

**Figure 16:** Λ-$[Fe(ox)_3]^{3-}$

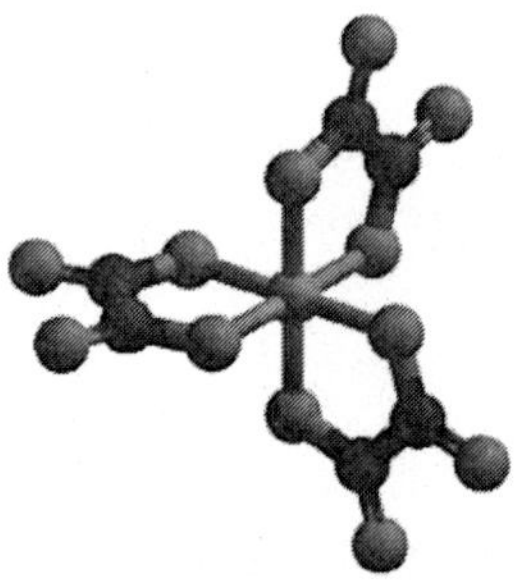

**Figure 17:** Δ-$[Fe(ox)_3]^{3-}$

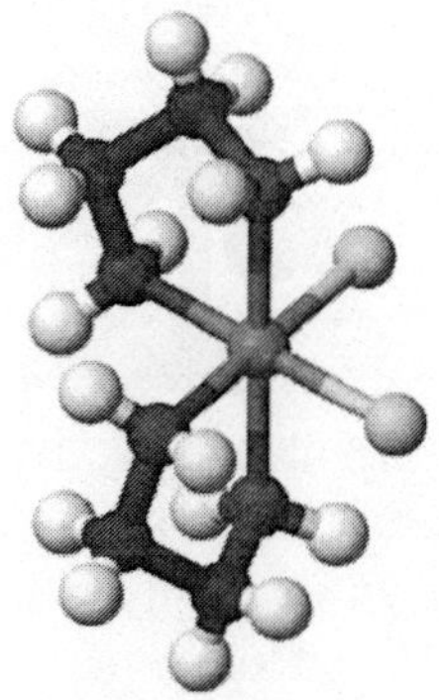

**Figure 18:** $\Lambda$-*cis*-$[CoCl_2(en)_2]^+$

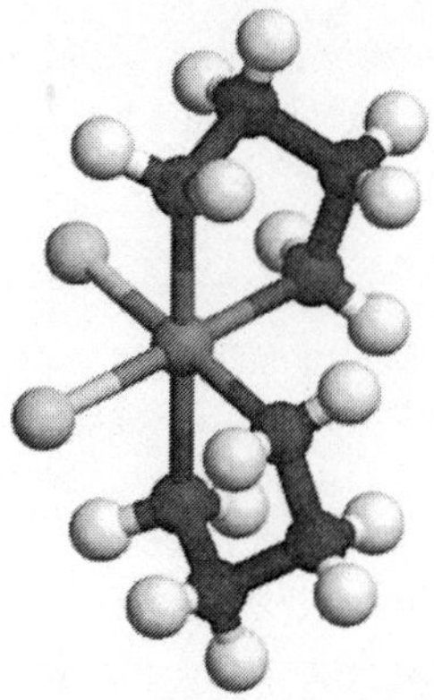

**Figure 19:** $\Delta$-*cis*-$[CoCl_2(en)_2]^+$

### *Structural Isomerism*

Structural isomerism occurs when the bonds are themselves different. Linkage isomerism is only one of several types of structural isomerism in coordination complexes (as well as other classes of chemical compounds). Linkage isomerism occurs with ambidentate ligands that can bind in more than one place. For example, $NO_2$ is an ambidentate ligand: It can bind to a metal at either the N atom or an O atom.

### *Electronic Properties*

Many of the properties of metal complexes are dictated by their electronic structures. The electronic structure can be described by a relatively ionic model that ascribes formal charges to the metals and ligands. This approach is the essence of crystal field theory (CFT). Crystal field theory, introduced by Hans Bethe in 1929, gives a quantum mechanically based attempt at understanding complexes. But crystal field theory treats all interactions in a complex as ionic and assumes that the ligands can be approximated by negative point charges.

More sophisticated models embrace covalency, and this approach is described by ligand field theory (LFT) and Molecular orbital theory (MO). Ligand field theory, introduced in 1935 and built from molecular orbital theory, can handle a broader range of complexes and can explain complexes in which the interactions are covalent. The chemical applications of group theory can aid in the understanding of crystal or ligand field theory, by allowing simple, symmetry based solutions to the formal equations.

Chemists tend to employ the simplest model required to predict the properties of interest; for this reason, CFT has been a favourite for the discussions when possible. MO and LF theories are more complicated, but provide a more realistic perspective.

### *Magnetism*

Metal complexes that have unpaired electrons are magnetic. Considering only monometallic complexes, unpaired electrons arise because the complex has an odd number of electrons or because electron pairing is destabilised. Thus, monomeric Ti(III) species have one "d-electron" and must be (para)magnetic, regardless of the geometry or the nature of the ligands. Ti(II), with two d-electrons, forms some complexes that have two unpaired electrons and others with none. This effect is illustrated by the compounds $TiX_2[(CH_3)_2PCH_2CH_2P(CH_3)_2]_2$: when X = Cl, the complex is paramagnetic (high-spin configuration), whereas when X=$CH_3$, it is diamagnetic (low-spin configuration). It is important to realise that ligands provide an important means of adjusting the ground state properties.

In bi- and polymetallic complexes, in which the individual centers have an odd number of electrons or that are high-spin, the situation is more complicated.

If there is interaction (either direct or through ligand) between the two (or more) metal centers, the electrons may couple (antiferromagnetic coupling, resulting in a diamagnetic compound), or they may enhance each other (ferromagnetic coupling). When there is no interaction, the two (or more) individual metal centers behave as if in two separate molecules.

### *Reactivity*

Complexes show a variety of possible reactivities:

- Electron transfers: A common reaction between coordination complexes involving ligands are inner and outer sphere electron transfers. They are two different mechanisms of electron transfer redox reactions, largely defined by the late Henry Taube. In an inner sphere reaction, a ligand with two lone electron pairs acts as a *bridging ligand*, a ligand to which both coordination centres can bond. Through this, electrons are transferred from one centre to another.
- (Degenerate) ligand exchange: One important indicator of reactivity is the rate of degenerate exchange of ligands. For example, the rate of interchange of coordinate water in $[M(H_2O)_6]^{n+}$ complexes varies over 20 orders of magnitude. Complexes where the ligands are released and rebound rapidly are classified as labile. Such labile complexes can be quite stable thermodynamically. Typical labile metal complexes either have low-charge ($Na^+$), electrons in d-orbitals that are antibonding with respect to the ligands ($Zn^{2+}$), or lack covalency ($Ln^{3+}$, where Ln is any lanthanide). The lability of a metal complex also depends on the high-spin vs. low-spin configurations when such is possible. Thus, high-spin Fe(II) and Co(III) form labile complexes, whereas low-spin analogues are inert. Cr(III) can exist only in the low-spin state (quartet), which is inert because of its high formal oxidation state, absence of electrons in orbitals that are M-L antibonding, plus some "ligand field stabilisation" associated with the d3 configuration.
- Associative processes: Complexes that have unfilled or half-filled orbitals often show the capability to react with substrates. Most substrates have a singlet ground-state; that is, they have lone electron pairs (e.g., water, amines, ethers), so these substrates need an empty orbital to be able to react with a metal centre. Some substrates (e.g., molecular oxygen) have a triplet ground state, which results that metals with half-filled orbitals have a tendency to react with such substrates (it must be said that the dioxygen-molecule also has lone pairs, so it is also capable to react as a 'normal' Lewis base).

If the ligands around the metal are carefully chosen, the metal can aid in (stoichiometric or catalytic) transformations of molecules or be used as a sensor.

### *Classification*

Metal complexes, also known as coordination compounds, include all metal compounds, aside from metal vapours, plasmas, and alloys. The study of "coordination chemistry" is the study of "inorganic chemistry" of all alkali and alkaline earth metals, transition metals, lanthanides, actinides, and metalloids. Thus, coordination chemistry is the chemistry of the majority of the periodic table. Metals and metal ions exist, in the condensed phases at least, only surrounded by ligands.

The areas of coordination chemistry can be classified according to the nature of the ligands, in broad terms:

- Classical (or "Werner Complexes"): Ligands in classical coordination chemistry bind to metals, almost exclusively, via their "lone pairs" of electrons residing on the main group atoms of the ligand. Typical ligands are $H_2O$, $NH_3$, $Cl^-$, $CN^-$,en$^-$

Examples: $[Co(EDTA)]^-$, $[Co(NH_3)_6]Cl_3$, $[Fe(C_2O_4)_3]K_3$

- Organometallic Chemistry: Ligands are organic (alkenes, alkynes, alkyls) as well as "organic-like" ligands such as phosphines, hydride, and CO.

Example: $(C_5H_5)Fe(CO)_2CH_3$

- Bioinorganic Chemistry: Ligands are those provided by nature, especially including the side chains of amino acids, and many cofactors such as porphyrins.

Example: hemoglobin

Many natural ligands are "classical" especially including water.

- Cluster Chemistry: Ligands are all of the above also include other metals as ligands.

Example $Ru_3(CO)_{12}$

- In some cases there are combinations of different fields:

Example: $[Fe_4S_4(Scysteinyl)_4]^{2"}$, in which a cluster is embedded in a biologically active species.

Mineralogy, materials science, and solid state chemistry - as they apply to metal ions - are subsets of coordination chemistry in the sense that the metals are surrounded by ligands. In many cases these ligands are oxides or sulfides, but the metals are coordinated nonetheless, and the principles and guidelines discussed below apply.

In hydrates, at least some of the ligands are water molecules. It is true that the focus of mineralogy, materials science, and solid state chemistry differs from the usual focus of coordination or inorganic chemistry.

The former are concerned primarily with polymeric structures, properties arising from a collective effects of many highly interconnected metals. In contrast, coordination chemistry focuses on reactivity and properties of complexes containing individual metal atoms or small ensembles of metal atoms.

### Older Classifications of Isomerism

Traditional classifications of the kinds of isomer have become archaic with the advent of modern structural chemistry. In the older literature, one encounters:

- Ionisation isomerism describes the possible isomers arising from the exchange between the outer sphere and inner sphere. This classification relies on an archaic classification of the inner and outer sphere. In this classification, the "outer sphere ligands," when ions in solution, may be switched with "inner sphere ligands" to produce an isomer.
- Solvation isomerism occurs when an inner sphere ligand is replaced by a solvent molecule. This classification is obsolete because it considers solvents as being distinct from other ligands. Some of the problems are discussed under water of crystallisation.

### Naming Complexes

The basic procedure for naming a complex:

- When naming a complex ion, the ligands are named before the metal ion.
- Write the names of the ligands in alphabetical order. (Numerical prefixes do not affect the order.)
  - o Multiple occurring monodentate ligands receive a prefix according to the number of occurrences: *di-*, *tri-*, *tetra-*, *penta-*, or *hexa*. Polydentate ligands (e.g., ethylenediamine, oxalate) receive *bis-*, *tris-*, *tetrakis-*, etc.
  - o Anions end in *ido*. This replaces the final 'e' when the anion ends with '-ate', e.g. *sulfate* becomes *sulfato*. It replaces 'ide': *cyanide* becomes *cyanido*.

- o Neutral ligands are given their usual name, with some exceptions: $NH_3$ becomes *ammine*; $H_2O$ becomes *aqua* or *aquo*; CO becomes *carbonyl*; NO becomes *nitrosyl.*
- Write the name of the central atom/ion. If the complex is an anion, the central atom's name will end in *-ate*, and its Latin name will be used if available (except for mercury).
- If the central atom's oxidation state needs to be specified (when it is one of several possible, or zero), write it as a Roman numeral (or 0) in parentheses.
- Name cation then anion as separate words (if applicable, as in last example)

Examples:

$[NiCl_4]^{2''}$ → tetrachloridonickelate(II) ion

$[CuNH_3Cl_5]^{3''}$ → amminepentachloridocuprate(II) ion

$[Cd(en)_2(CN)_2]$ → dicyanidobis(ethylenediamine)cadmium(II)

$[Co(NH_3)_5Cl]SO_4$ → pentaamminechloridocobalt(III) sulfate

The coordination number of ligands attached to more than one metal (bridging ligands) is indicated by a subscript to the Greek symbol μ placed before the ligand name. Thus the dimer of aluminium trichloride is described by $Al_2Cl_4(\mu_2\text{-}Cl)_2$.

### *Application of Coordination Compounds*

1. They are used in photography, i.e., AgBr forms a soluble complex with sodium thiosulfate in photography.
2. $K[Ag(CN)_2]$ is used for electroplating of silver, and $K[Au(CN)_2]$is used for gold plating.
3. Some ligands oxidise $Co^{2+}$ to $Co^{3+}$ ion.
4. Ethylenediaminetetraacetic acid (EDTA) is used for estimation of $Ca^{2+}$ and $Mg^{2+}$in hard water.
5. Silver and gold are extracted by treating zinc with their cyanide complexes.

## Coordination Geometry

The term coordination geometry is used in a number of related fields of chemistry and solid state chemistry/physics.

### *Molecules*

The coordination geometry of an atom is the geometrical pattern formed by atoms around the central atom.

### *Inorganic Coordination Complexes*

In the field of inorganic coordination complexes it is the geometrical pattern formed by the atoms in the ligands that are bonded to the central atom in a molecule or a coordination complex.

The geometrical arrangement will vary according to the number and type of ligands bonded to the metal centre, and to the coordination preference of the central atom, typically a metal in a coordination complex.

The number of atoms bonded, (i.e. the number of Σ-bonds between central atom and ligands) is termed the coordination number. The geometrical pattern can be described as a polyhedron where the vertices of the polyhedron are the centres of the coordinating atoms in the ligands.

The coordination preference of a metal often varies with its oxidation state. The number of coordination bonds (coordination number) can vary from two as high as 20 in $Th(ç^5\text{-}C_5H_5)_4$.

One of the most common coordination geometries is octahedral, where six ligands are coordinated to the metal in a symmetrical distribution, leading to the formation of an octahedron if lines were drawn between the ligands. Other common coordination geometries are tetrahedral and square planar.

Crystal field theory may be used to explain the relative stabilities of transition metal compounds of different coordination geometry, as well as the presence or absence of paramagnetism.

VSEPR may be used for complexes of main group element to predict geometry.

### *Crystallography Usage*

In a crystal structure the coordination geometry of an atom is the geometrical pattern of coordinating atoms where the definition of coordinating atoms depends on the bonding model used. For example in the rock salt ionic structure each sodium atom has six near neighbour chloride ions in an octahedral geometry and each chloride has similarly six near neighbour sodium ions in an octahedral geometry. In metals with the body centred cubic (bcc) structure each atom has eight nearest neighbours in a cubic geometry.

In metals with the face centred cubic (fcc) structure each atom has twelve nearest neighbours in a cuboctahedral geometry.

### *Naming of Inorganic Compounds*

IUPAC have introduced the polyhedral symbol as part of their IUPAC nomenclature of inorganic chemistry 2005 recommendations to describe the geometry around an atom in a compound.

IUCr have proposed a symbol which is shown as a superscript in square brackets in the chemical formula. For example $CaF_2$ would be $Ca^{[8cb]}F_2^{[4t]}$, where [8cb] means cubic coordination and [4t] means tetrahedral. The equivalent symbols in IUPAC are *CU*"8 and *T*"4 respectively.

The IUPAC symbol is applicable to complexes and molecules whereas the IUCr proposal applies to crystalline solids.

# Chapter 3

# Organometallic Chemistry

Organometallic chemistry is the study of chemical compounds containing bonds between carbon and a metal. Since many compounds without such bonds are chemically similar, an alternative may be compounds containing metal-element bonds of a largely covalent character. Organometallic chemistry combines aspects of inorganic chemistry and organic chemistry.

## Organometallic Compounds

Organometallic compounds are distinguished by the prefix "organo-" e.g. organopalladium compounds. Examples of such organometallic compounds include all Gilman reagents, which contain lithium and copper. Tetracarbonyl nickel, and ferrocene are examples of organometallic compounds containing transition metals. Other examples include organomagnesium compounds like iodo (methyl) magnesium MeMgI, diethylmagnesium ($Et_2Mg$), and all Grignard reagents; organolithium compounds such as n-butyllithium (n-BuLi), organozinc compounds such as diethylzinc ($Et_2Zn$) and chloro(ethoxycarbonylmethyl)zinc ($ClZnCH_2C(=O)OEt$); and organocopper compounds such as lithium dimethylcuprate ($Li^+[CuMe_2]^-$).

The term "metalorganics" usually refers to metal-containing compounds lacking direct metal-carbon bonds but which contain organic ligands. Metal beta-diketonates, alkoxides, and dialkylamides are representative members of this class. In addition to the traditional metals, lanthanides, actinides, and semimetals, elements such as boron, silicon, arsenic, and selenium are considered to form organometallic compounds, e.g. organoborane compounds such as triethylborane ($Et_3B$).

## Coordination Compounds with Organic Ligands

Many complexes feature coordination bonds between a metal and organic ligands. The organic ligands often bind the metal through a heteroatom such as oxygen or nitrogen, in which case such compounds are considered coordination compounds. However, if any of the ligands form a direct M-C bond, then complex is usually considered to be organometallic, e.g., $[(C_6H_6)Ru(H_2O)_3]^{2+}$. Furthermore, many lipophilic compounds such as metal acetylacetonates and metal alkoxides are called "metalorganics."

Many organic coordination compounds occur naturally. For example, hemoglobin and myoglobin contain an iron center coordinated to the nitrogen atoms of a porphyrin ring; magnesium is the center of a chlorin ring in chlorophyll. The field of such inorganic compounds is known as bioinorganic chemistry. In contrast to these coordination compounds, methylcobalamin (a form of Vitamin $B_{12}$), with a cobalt-methyl bond, is a true organometallic complex, one of the few known in biology. This subset of complexes are often discussed within the subfield of bioorganometallic chemistry. Illustrative of the many functions of the $B_{12}$-dependent enzymes, the MTR enzyme catalyses the transfer of a methyl group from a nitrogen on N5-methyl-tetrahydrofolate to the sulfur of homocysteine to produce methionine.

The status of compounds in which the canonical anion has a delocalised structure in which the negative charge is shared with an atom more electronegative than carbon, as in enolates, may vary with the nature of the anionic moiety, the metal ion, and possibly the medium; in the absence of direct structural evidence for a carbon–metal bond, such compounds are not considered to be organometallic.

## Structure and Properties

The metal-carbon bond in organometallic compounds is generally of character intermediate between ionic and covalent. Primarily ionic metal-carbon bonds are encountered either when the metal is very electropositive (as in the case of Group 1 or Group 2 metals) or when the carbon-containing ligand exists as a stable carbanion. Carbanions can be stabilised by resonance (as in the case of the aromatic cyclopentadienyl anion) or by the presence of electron-withdrawing substituents (as in the case of the triphenylmethyl anion). Hence, the bonding in compounds like sodium acetylide and triphenylmethyl potassium is primarily ionic. On the other hand, the ionic character of metal-carbon bonds in the organometallic compounds of transition metals, poor metals, and metalloids tends to be intermediate, owing to the middle-of-the-road electronegativity of such metals. Organometallic compounds with bonds that have characters in between ionic and covalent are very important in industry, as they are both relatively stable in solutions and relatively ionic to undergo reactions. Two important classes are organolithium and Grignard reagents. In certain organometallic compounds such as ferrocene or dibenzenechromium, the pi orbitals of the organic moiety ligate the metal.

## Applications

Organometallics find practical uses in stoichiometric and catalytic processes, especially processes involving carbon monoxide and alkene-derived polymers. All the world's polyethylene and polypropylene are produced via organometallic catalysts, usually heterogeneously via Ziegler-Natta catalysis. Acetic acid is produced via metal carbonyl catalysts in the Monsanto process and Cativa process. Most synthetic aldehydes are produced via hydroformylation. The bulk of the synthetic alcohols, at least those larger than ethanol, are produced by hydrogenation of hydroformylation derived aldehydes. Similarly, the Wacker process is used in the oxidation of ethylene to acetaldehyde.

Organolithium, organomagnesium, and organoaluminium compounds are highly basic and highly reducing. They catalyse many polymerisation reactions, but are also useful stoichiometrically.

III-V semiconductors are produced from trimethylgallium, trimethylindium, trimethylaluminum and related nitrogen / phosphorus / arsenic / antimony compounds. These volatile compounds are decomposed along with ammonia, arsine, phosphine and related hydrides on a heated substrate via metalorganic vapour phase epitaxy (MOVPE) process for applications such as light emitting diodes (LEDs) fabrication.

Organometallic compounds may be found in the environment and some, such as organo-lead and organo mercury compounds are a toxic hazard.

## Concepts and Techniques

As in other areas of chemistry, electron counting is useful for organising organometallic chemistry. The 18-electron rule is helpful in predicting the stabilities of metal carbonyls and related compounds. Chemical bonding and reactivity in organometallic compounds is often discussed from the perspective of the isolobal principle.

NMR and infrared spectroscopy are common techniques used to determine structure. The dynamic properties of organometallic compounds is often probed with variable-temperature NMR and chemical kinetics.

Organometallic compounds undergo several important reactions:

- oxidative addition and reductive elimination
- transmetalation
- carbometalation
- Hydrometalation
- electron transfer
- beta-hydride elimination
- organometallic substitution reaction
- carbon-hydrogen bond activation
- cyclometalation
- Migratory insertion.

### *History*

Early developments in organometallic chemistry include Louis Claude Cadet's synthesis of methyl arsenic compounds related to cacodyl, William Christopher Zeise's platinum-ethylene complex, Edward Frankland's discovery of dimethyl zinc, Ludwig Mond's discovery of $Ni(CO)_4$, and Victor Grignard's organomagnesium compounds. The abundant and diverse products from coal and petroleum led to Ziegler-Natta, Fischer-Tropsch, hydroformylation catalysis which employ CO, $H_2$, and alkenes as feedstocks and ligands.

Tetraethyllead previously was combined with gasoline as an antiknock agent. Due to lead's toxicity it is no longer used, its replacements being other organometallic compounds such as ferrocene and methylcyclopentadienyl manganese tricarbonyl (MMT). Recognition of organometallic chemistry as a distinct subfield culminated in the Nobel Prizes to Ernst Fischer and Geoffrey Wilkinson for work on metallocenes. In 2005, Yves Chauvin, Robert H. Grubbs and Richard R. Schrock shared the Nobel Prize for metal-catalysed olefin metathesis.

## Chelation

Chelation is the formation or presence of two or more separate coordinate bonds between a polydentate (multiple bonded) ligand and a single central atom. Usually these ligands are organic compounds, and are called chelants, chelators, chelating agents, or sequestering agents. The ligand forms a chelate complex with the substrate. Chelate complexes are contrasted with coordination complexes composed of monodentate ligands, which form only one bond with the central atom.

Chelants, according to ASTM-A-380, are "chemicals that form soluble, complex molecules with certain metal ions, inactivating the ions so that they cannot normally react with other elements or ions to produce precipitates or scale."

The word chelation is derived from Greek ÷çλÞ, *chelè*, meaning claw; the ligands lie around the central atom like the claws of a lobster.

### *The Chelate Effect*

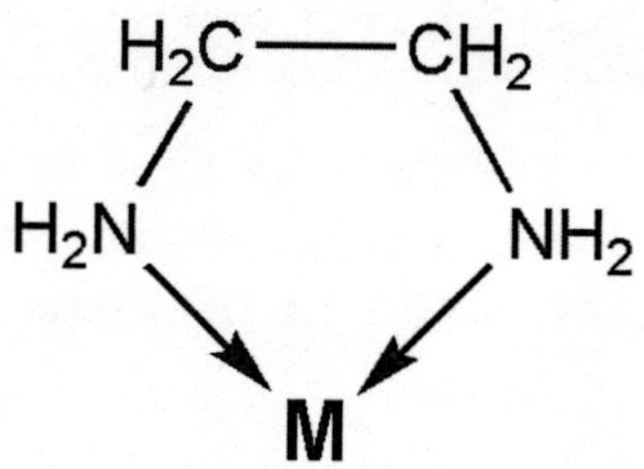


**Figure 1:** Ethylenediamine ligand, binding to a central metal ion with two bonds

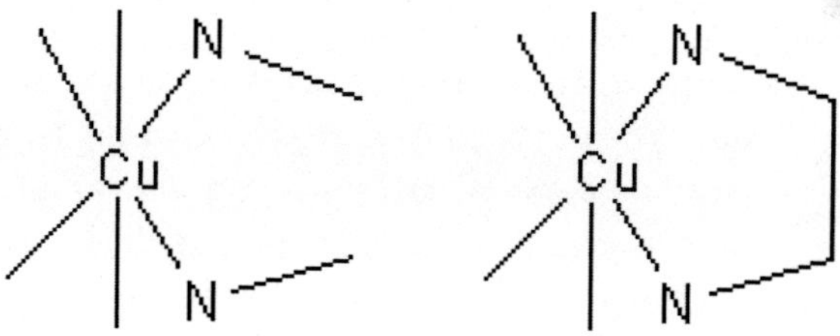


**Figure 2:** $Cu^{2+}$ complexes with methylamine (left)*and* ethylenediamine (right)

The chelate effect describes the enhanced affinity of chelating ligands for a metal ion compared to the affinity of a collection of similar nonchelating (monodentate) ligands for the same metal.

Consider the two equilibria, in aqueous solution, between the copper(II) ion, $Cu^{2+}$ and ethylenediamine (en) on the one hand and methylamine, $MeNH_2$ on the other.

$$Cu^{2+} + en \rightleftharpoons [Cu(en)]^{2+} (1) \quad (1)$$

$$Cu^{2+} + 2\ MeNH_2\ \ [Cu(MeNH_2)_2]^{2+} \quad (2)$$

In (1) the bidentate ligand ethylene diamine forms a chelate complex with the copper ion. Chelation results in the formation of a five–membered ring. In (2) the bidentate ligand is replaced by two monodentate methylamine ligands of approximately the same donor power, meaning that the enthalpy of formation of Cu—N bonds is approximately the same in the two reactions. Under conditions of equal copper concentrations and when the concentration of methylamine is twice the concentration of ethylenediamine, the concentration of the complex (1) will be greater than the concentration of the complex (2). The effect increases with the number of chelate rings so the concentration of the EDTA complex, which has six chelate rings, is much higher than a corresponding complex with two monodentate nitrogen donor ligands and four monodentate carboxylate ligands. Thus, the phenomenon of the chelate effect is a firmly established empirical fact.

The thermodynamic approach to explaining the chelate effect considers the equilibrium constant for the reaction: the larger the equilibrium constant, the higher the concentration of the complex.

$$[Cu(en)] = \beta_{11}[Cu][en]$$

$$[Cu(MeNH_2)_2] = \beta_{12}[Cu][MeNH_2]^2$$

Electrical charges have been omitted for simplicity of notation. The square brackets indicate concentration, and the subscripts to the stability constants, a, indicate the stoichiometry of the complex. When the analytical concentration of methylamine is twice that of ethylenediamine and the concentration of copper is the same in both reactions, the concentration [Cu(en)] is much higher than the concentration $[Cu(MeNH_2)_2]$ because $\beta_{11} >> \beta_{12}$.

An equilibrium constant, *K*, is related to the standard Gibbs free energy, $\Delta G$ by

$$\Delta G = \text{“}RT \ln K = \Delta H \cdot T\Delta S$$

where *R* is the gas constant and *T* is the temperature in kelvins. $\Delta H$ is the standard enthalpy change of the reaction and $\Delta S$ is the standard entropy change.

It has already been posited that the enthalpy term should be approximately the same for the two reactions. Therefore the difference between the two stability constants is due to the entropy term. In equation (1) there are two particles on the left and one on the right, whereas in equation (2) there are three particles on the left and one on the right. This means that less entropy of disorder is lost when the chelate complex is formed than when the complex with monodentate ligands is formed. This is one of the factors contributing to the entropy difference. Other factors include solvation changes and ring formation. Some experimental data to illustrate the effect are shown in the following table.

| Equilibrium | log β | $\Delta G^{\ominus}$ | $\Delta H^{\ominus}$ /kJ mol$^{-1}$ | $-T\Delta S^{\ominus}$ /kJ mol$^{-1}$ |
|---|---|---|---|---|
| $Cd^{2+} + 4\ MeNH_2 \rightleftharpoons Cd(MeNH_2)_4^{2+}$ | 6.55 | -37.4 | -57.3 | 19.9 |
| $Cd^{2+} + 2\ en \rightleftharpoons Cd(en)_2^{2+}$ | 10.62 | -60.67 | -56.48 | -4.19 |

These data show that the standard enthalpy changes are indeed approximately equal for the two reactions and that the main reason why that the greater stability of the chelate complex is due to the entropy term, which is much less unfavourable, indeed, it is favourable in this instance. In general it is difficult to account precisely for thermodynamic values in terms of changes in solution at the molecular level, but it is clear that the chelate effect is predominantly an effect of entropy.

Other explanations, Including that of Schwarzenbach, are discussed in Greenwood and Earnshaw (*loc.cit*).

### *In Nature*

Virtually all biochemicals exhibit the ability to dissolve certain metal cations. Thus, proteins, polysaccharides, and polynucleic acids are excellent polydentate ligands for many metal ions. Organic compounds such as the amino acids glycine and histidine, organic diacids such as malate, and polypeptides such as phytochelatin are also typical chelators. In addition to these adventitious chelators, several biomolecules are specifically produced to bind certain metals.

### *In Biochemistry and Microbiology*

Virtually all metalloenzymes feature metals that are chelated, usually to peptides or cofactors and prosthetic groups. Such chelating agents include the porphyrin rings in hemoglobin and chlorophyll. Many microbial species produce water-soluble pigments that serve as chelating agents, termed siderophores. For example, species of *Pseudomonas* are known to secrete pyocyanin and pyoverdin that bind iron. Enterobactin, produced by E. coli, is the strongest chelating agent known.

### In Geology

In earth science, chemical weathering is attributed to organic chelating agents, *e.g.* peptides and sugars, that extract metal ions from minerals and rocks. Most metal complexes in the environment and in nature are bound in some form of chelate ring, *e.g.* with a humic acid or a protein. Thus, metal chelates are relevant to the mobilisation of metals in the soil, the uptake and the accumulation of metals into plants and micro-organisms. Selective chelation of heavy metals is relevant to bioremediation, *e.g.* removal of $^{137}Cs$ from radioactive waste.

### Applications

Chelators are used in producing nutritional supplements, fertilizers, chemical analysis, as water softeners, commercial products such as shampoos and food preservatives, medicine, heavy metal detox, and industrial applications.

### Nutritional Supplements

In the 1960's, scientists developed the concept of chelating a metal ion prior to feeding the element to the animal. They believed that this would create a neutral compound, protecting the mineral from being complexed with insoluble salts within the stomach, rendering the metal unavailable for absorption. Amino acids, being effective metal binders, were chosen as the prospective ligands, and research was conducted on the metal-amino acid combinations. The research supported that the metal-amino acid chelates were able to enhance mineral absorption.

During this period, synthetic chelates were also being developed. An example of such synthetics is ethylenediaminetetraacetic acid (EDTA). These synthetics applied the same concept of chelation and did create chelated compounds; however, these synthetics were too stable and not nutritionally viable. If the mineral was taken from the EDTA ligand, the ligand could not be used by the body and would be expelled. During the expulsion process the EDTA ligand will randomly chelate and strip another mineral from the body.

According to the Association of American Feed Control Officials (AAFCO), a metal amino acid chelate is defined as the product resulting from the reaction of a metal ion from a soluble metal salt with a mole ratio of one to three (preferably two) moles of amino acids. The average weight of the hydrolysed amino acids must be approximately 150 and the resulting molecular weight of the chelate must not exceed 800 Da.

Since the early development of these compounds, much more research has been conducted, and has been applied to human nutrition products in a similar manner to the animal nutrition experiments that pioneered the technology. Ferrous bis-glycinate is an example of one of these compounds that has been developed for human nutrition.

### *Fertilizers*

Many mineral deficiencies can occur in plants, such as iron chlorosis, which can reduce the nutritional benefits of crops and eventually result in plant death. Mineral chelates have been used to alleviate the mineral deficiencies of affected crops through liquid foliar applications. These fertilizers are also used to prevent deficiencies from occurring and improving the overall health of the plants.

### *Heavy Metal Detoxification*

Chelation therapy is the use of chelating agents to detoxify poisonous metal agents such as mercury, arsenic, and lead by converting them to a chemically inert form that can be excreted without further interaction with the body, and was approved by the U.S. Food and Drug Administration in 1991. In alternative medicine, chelation is used as a treatment for autism, although this practice is controversial due to the absence of scientific plausibility, lack of FDA approval, and its potentially deadly side-effects.

Although they can be beneficial in cases of heavy metal poisoning, chelating agents can also be dangerous. Use of disodium EDTA instead of calcium EDTA has resulted in fatalities due to hypocalcemia.

### *Other Medical Applications*

Antibiotic drugs of the tetracycline family are chelators of $Ca^{2+}$ and $Mg^{2+}$ ions.

EDTA is also used in root canal treatment as a way to irrigate the canal. EDTA softens the dentin facilitating access to the entire canal length and to remove the smear layer formed during instrumentation.

Chelate complexes of gadolinium are often used as contrast agents in MRI scans.

### *Chemical Applications*

Homogeneous catalysts are often chelated complexes. A typical example is the ruthenium(II) chloride chelated with BINAP (a bidentate phosphine) used in e.g. Noyori asymmetric hydrogenation and asymmetric isomerization. The latter has the practical use of manufacture of synthetic (–)-menthol.

Citric acid is used to soften water in soaps and laundry detergents. A common synthetic chelator is EDTA. Phosphonates are also well known chelating agents. Chelators are used in water treatment programs and specifically in steam engineering, e.g., boiler water treatment system: *Chelant Water Treatment system.*

Products such as Evapo-Rust are chelating agents sold for the removal of rust from iron and steel.

## Van der Waals Force

In physical chemistry, the van der Waals force (or van der Waals interaction), named after Dutch scientist Johannes Diderik van der Waals, is the sum of the attractive or repulsive forces between molecules (or between parts of the same molecule) other than those due to covalent bonds or to the electrostatic interaction of ions with one another or with neutral molecules. The term includes:

- force between two permanent dipoles (Keesom force)
- force between a permanent dipole and a corresponding induced dipole (Debye force)
- force between two instantaneously induced dipoles (London dispersion force).

It is also sometimes used loosely as a synonym for the totality of intermolecular forces. Van der Waals forces are relatively weak compared to normal chemical bonds, but play a fundamental role in fields as diverse as supramolecular chemistry, structural biology, polymer science, nanotechnology, surface science, and condensed matter physics. Van der Waals forces define the chemical character of many organic compounds. They also define the solubility of organic substances in polar and non-polar media. In low molecular weight alcohols, the properties of the polar hydroxyl group dominate the weak intermolecular forces of van der Waals. In higher molecular weight alcohols, the properties of the nonpolar hydrocarbon chain(s) dominate and define the solubility. Van der Waals-London forces grow with the length of the nonpolar part of the substance.

### *Definition*

Van der Waals forces include attractions between atoms, molecules, and surfaces, as well as other intermolecular forces. They differ from covalent and ionic bonding in that they are caused by correlations in the fluctuating polarisations of nearby particles (a consequence of quantum dynamics).

Intermolecular forces have four major contributions:

1. A repulsive component resulting from the Pauli exclusion principle that prevents the collapse of molecules.
2. Attractive or repulsive electrostatic interactions between permanent charges (in the case of molecular ions), dipoles (in the case of molecules without inversion center), quadrupoles (all molecules with symmetry lower than cubic), and in general between permanent multipoles. The electrostatic interaction is sometimes called the Keesom interaction or Keesom force after Willem Hendrik Keesom.
3. Induction (also known as polarisation), which is the attractive interaction between a permanent multipole on one molecule with an induced multipole on another. This interaction is sometimes called Debye force after Peter J.W. Debye.
4. Dispersion (usually named after Fritz London), which is the attractive interaction between any pair of molecules, including non-polar atoms, arising from the interactions of instantaneous multipoles.

Returning to nomenclature, different texts refer to different things using the term "van der Waals force". Some texts mean by the van der Waals force the totality of forces (including repulsion); others mean all the attractive forces (and then sometimes distinguish van der Waals-Keesom, van der Waals-Debye, and van der Waals-London).

All intermolecular/van der Waals forces are anisotropic (except those between two noble gas atoms), which means that they depend on the relative orientation of the molecules. The induction and dispersion interactions are always attractive, irrespective of orientation, but the electrostatic interaction changes sign upon rotation of the molecules. That is, the electrostatic force can be attractive or repulsive, depending on the mutual orientation of the molecules. When molecules are in thermal motion, as they are in the gas and liquid phase, the electrostatic force is averaged out to a large extent, because the molecules thermally rotate and thus probe both repulsive and attractive parts of the electrostatic force. Sometimes this effect is expressed by the statement that "random thermal motion around room temperature can usually overcome or disrupt them" (which refers to the electrostatic component of the van der Waals force). Clearly, the thermal averaging effect is much less pronounced for the attractive induction and dispersion forces.

The Lennard-Jones potential is often used as an approximate model for the isotropic part of a total (repulsion plus attraction) van der Waals force as a function of distance.

Van der Waals forces are responsible for certain cases of pressure broadening (van der Waals broadening) of spectral lines and the formation of van der Waals molecules. The London-van der Waals forces are related to the Casimir effect for dielectric media, the former being the microscopic description of the latter bulk property. The first detailed calculations of this were done in 1955 by E. M. Lifshitz.

### London Dispersion Force

London dispersion forces, named after the German-American physicist Fritz London, are weak intermolecular forces that arise from the interactive forces between instantaneous multipoles in molecules without permanent multipole moments. These forces dominate the interaction of non-polar molecules, and also play a less significant role in Van der Waals forces than molecules containing permanent dipoles or ionized molecules. London dispersion forces are also known as dispersion forces, London forces, or instantaneous dipole–induced dipole forces. They increase with the molar mass, causing a higher boiling point especially for the halogen group.

### Van der Waals Forces between Macroscopic Objects

For macroscopic bodies with known volumes and numbers of atoms or molecules per unit volume, the total van der Waals force is often computed based on the "microscopic theory" as the sum over all interacting pairs. It is necessary to integrate over the total volume of the object, which makes the calculation dependent on the objects' shapes. For example, the van der Waals interaction energy between spherical bodies of radii $R_1$ and $R_2$ and with smooth surfaces was approximated in 1937 by Hamaker (using London's famous 1937 equation for the dispersion interaction energy between atoms/ molecules as the starting point) by:

$$U(z;R_1,R_2)=-\frac{A}{6}\left[\frac{2R_1R_2}{z^2-(R_1+R_2)^2}+\frac{2R_1R_2}{z^2-(R_1-R_2)^2}+\ln\left(\frac{z^2-(R_1+R_2)^2}{z^2-(R_1-R_2)^2}\right)\right] \quad (1)$$

where A is the Hamaker coefficient, which is a constant (~10"19 - 10"20 J) that depends on the material properties (it can be positive or negative in sign depending on the intervening medium), and $z$ is the center-to-center distance, i.e. the sum of $R_1$, $R_2$, and $r$ (the distance between the surfaces):

$$z=R_1+R_2+r\,.$$

In the limit of close-approach, the spheres are sufficiently large compared to the distance between them, i.e. $r \ll R_1 \text{or} R_2$, so that equation (1) for the potential energy function simplifies to:

$$U(r;R_1,R_2)=-\frac{AR_1R_2}{(R_1+R_2)6r} \quad (2)$$

The van der Waals *force* between two spheres of constant radii ($R_1$ and $R_2$ are treated as parameters) is then a function of separation since the force on an object is the negative of the derivative of the potential energy function, $F_{VW}(r)=-\frac{dU(r)}{dr}$. This yields:

$$F_{VW}(r)=-\frac{AR_1R_2}{(R_1+R_2)6r^2} \quad (3)$$

The van der Waals forces between objects with other geometries using the Hamaker model have been published in the literature.

From the expression above, it is seen that the van der Waals force decreases with decreasing particle size (R). Nevertheless, the strength of inertial forces, such as gravity and drag/lift, decrease to a greater extent. Consequently, the van der Waals forces become dominant for collections of very small particles such as very fine-grained dry powders (where there are no capillary forces present) even though the force of attraction is smaller in magnitude than it is for larger particles of the same substance. Such powders are said to be cohesive, meaning they are not as easily fluidized or pneumatically conveyed as easily as their more coarse-grained counterparts. Generally, free-flow occurs with particles greater than about 250 μm.

The van der Waals force of adhesion is also dependent on the surface topography. If there are surface asperities, or protuberances, that result in a greater total area of contact between two particles or between a particle and a wall, this increases the van der Waals force of attraction as well as the tendency for mechanical interlocking.

The microscopic theory assumes pairwise additivity. It neglects many-body interactions and retardation. A more rigorous approach accounting for these effects, called the "macroscopic theory," was developed by Lifshitz in 1956. Langbein derived a much more cumbersome "exact" expression in 1970 for spherical bodies within the framework of the Lifshitz theory while a simpler macroscopic model approximation had been made by Derjaguin as early as 1934. Expressions for the van der Waals forces for many different geometries using the Lifshitz theory have likewise been published.

### *Use by Animals*

The ability of geckos – which can hang on a glass surface using only one toe – to climb on sheer surfaces has been attributed to the van der Waals forces between these surfaces and the spatula (plural spatulae), or microscopic projections, which cover the hair-like setae found on their footpads. A later study suggested that capillary adhesion might play a role, but that hypothesis has been rejected by more recent studies. There were efforts in 2008 to create a dry glue that exploits the effect, and success was achieved in 2011 to create a adhesive tape on similar grounds. In 2011, a paper was published relating the effect to both velcro-like hairs and the presence of lipids in gecko footprints.

## Electrostatics

Electrostatics is the branch of physics that deals with the phenomena and properties of stationary or slow-moving (without acceleration) electric charges.

Since classical antiquity, it was known that some materials such as amber attract lightweight particles after rubbing. The Greek word for amber, Þλεêτñïí *electron*, was the source of the word 'electricity'. Electrostatic phenomena arise from the forces that electric charges exert on each other. Such forces are described by Coulomb's law. Even though electrostatically induced forces seem to be rather weak, the electrostatic force between e.g. an electron and a proton, that together make up a hydrogen atom, is about 40 orders of magnitude stronger than the gravitational force acting between them. Electrostatic phenomena include many examples as simple as the attraction of the plastic wrap to your hand after you remove it from a package, to the apparently spontaneous explosion of grain silos, to damage of electronic components during manufacturing, to the operation of photocopiers.

Electrostatics involves the buildup of charge on the surface of objects due to contact with other surfaces. Although charge exchange happens whenever any two surfaces contact and separate, the effects of charge exchange are usually only noticed when at least one of the surfaces has a high resistance to electrical flow. This is because the charges that transfer to or from the highly resistive surface are more or less trapped there for a long enough time for their effects to be observed. These charges then remain on the object until they either bleed off to ground or are quickly neutralised by a discharge: e.g., the familiar phenomenon of a static 'shock' is caused by the neutralisation of charge built up in the body from contact with nonconductive surfaces.

## Fundamental Concepts

***Coulomb's Law:*** The fundamental equation of electrostatics is Coulomb's law, which describes the force between two point charges. The magnitude of the electrostatic force between two point electric charges $Q_1$ and $Q_2$ is directly proportional to the product of the magnitudes of each charge and inversely proportional to the surface area of a sphere whose radius is equal to the distance between the charges:

$$F = \frac{Q_1 Q_2}{4\pi r^2 \varepsilon_0},$$

where $\varepsilon_0$ is a constant called the permittivity of free space, a defined value:

$$\varepsilon_0 \overset{\text{def}}{=} \frac{1}{\mu_0 c_0^{\ 2}} = 8.854\,187\,817 \times 10^{-12} \quad ]$$

in $A^2s^4\ kg^{-1}m^{-3}$ or $C^2N^{-1}m^{-2}$ or $F\ m^{-1}$.

## The Electric Field

The electric field (in units of volts per metre) at a point is defined as the force (in newtons) per unit charge (in coulombs) on a charge at that point:

$$\vec{E} = \frac{\vec{F}}{q}.$$

From this definition and Coulomb's law, it follows that the magnitude of the electric field $E$ created by a single point charge $Q$ is:

$$E(\vec{r}) = \frac{Q}{4\pi r^2 \varepsilon_0}.$$

The electric field produced by a distribution of charges given by the volume charge density $\rho(\vec{r})$ is obtained by a triple integral of a vector function:

$$\vec{E}(\vec{r}) = \frac{1}{4\pi\epsilon_0} \iiint \frac{\vec{r} - \vec{r}'}{\|\vec{r} - \vec{r}'\|^3} \rho(\vec{r}') d^3 r'.$$

### *Gauss's Law*

Gauss' law states that "the total electric flux through a closed surface is proportional to the total electric charge enclosed within the surface".

Mathematically, Gauss's law takes the form of an integral equation:

$$\oint_S \vec{E} \cdot d\vec{A} = \int_V \frac{\rho}{\varepsilon_0} \cdot dV.$$

Alternatively, in differential form, the equation becomes

$$\vec{\nabla} \cdot \vec{E} = \frac{\rho}{\varepsilon_0}.$$

where $\vec{\nabla}$ is the divergence operator.

### *Poisson's Equation*

The definition of electrostatic potential, combined with the differential form of Gauss's law (above), provides a relationship between the potential $\phi$ and the charge density ñ:

$$\nabla^2 \phi = -\frac{\rho}{\varepsilon_0}.$$

This relationship is a form of Poisson's equation. Where $\varepsilon_0$ is vacuum permittivity.

### *Laplace's Equation*

In the absence of unpaired electric charge, the equation becomes

$$\vec{\nabla} \times \vec{E} = 0.$$

which is Laplace's equation.

### *The Electrostatic Approximation*

The validity of the electrostatic approximation rests on the assumption that the electric field is irrotational:

$$\vec{\nabla}\times\vec{E}=0.$$

From Faraday's law, this assumption implies the absence or near-absence of time-varying magnetic fields:

$$\frac{\partial\vec{B}}{\partial t}=0.$$

In other words, electrostatics does not require the absence of magnetic fields or electric currents. Rather, if magnetic fields or electric currents *do* exist, they must not change with time, or in the worst-case, they must change with time only *very slowly*. In some problems, both electrostatics and magnetostatics may be required for accurate predictions, but the coupling between the two can still be ignored.

### *Electrostatic Potential*

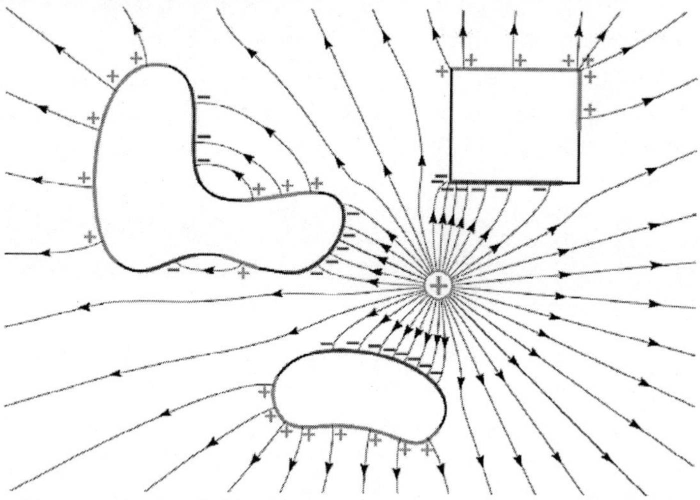

The electrostatic field *(lines with arrows)* of a nearby positive charge *(+)* causes the mobile charges in conductive objects to separate due to electrostatic induction. Negative charges are attracted and move to the surface of the object facing the external charge. Positive charges are repelled and move to the surface facing away. These induced surface charges are exactly the right size and shape so their opposing electric field cancels the electric field of the external charge throughout the interior of the metal. Therefore the electrostatic field everywhere inside a conductive object is zero, and the electrostatic potential is constant.

Because the electric field is irrotational, it is possible to express the electric field as the gradient of a scalar function, called the electrostatic potential (also known as the voltage). An electric field, *E*, points from regions of high potential, $\phi$, to regions of low potential, expressed mathematically as

$$\vec{E} = -\vec{\nabla}\phi.$$

The electrostatic potential at a point can be defined as the amount of work per unit charge required to move a charge from infinity to the given point.

### Triboelectric Series

The triboelectric effect is a type of contact electrification in which certain materials become electrically charged when they are brought into contact with a different material and then separated. One of the materials acquires a positive charge, and the other acquires an equal negative charge. The polarity and strength of the charges produced differ according to the materials, surface roughness, temperature, strain, and other properties. Amber, for example, can acquire an electric charge by friction with a material like wool. This property, first recorded by Thales of Miletus, was the first electrical phenomenon investigated by man. Other examples of materials that can acquire a significant charge when rubbed together include glass rubbed with silk, and hard rubber rubbed with fur.

### Electrostatic Generators

The presence of surface charge imbalance means that the objects will exhibit attractive or repulsive forces. This surface charge imbalance, which yields static electricity, can be generated by touching two differing surfaces together and then separating them due to the phenomena of contact electrification and the triboelectric effect. Rubbing two nonconductive objects generates a great amount of static electricity. This is not just the result of friction; two nonconductive surfaces can become charged by just being placed one on top of the other. Since most surfaces have a rough texture, it takes longer to achieve charging through contact than through rubbing. Rubbing objects together increases amount of adhesive contact between the two surfaces. Usually insulators, e.g., substances that do not conduct electricity, are good at both generating, and holding, a surface charge. Some examples of these substances are rubber, plastic, glass, and pith. Conductive objects only rarely generate charge imbalance except, for example, when a metal surface is impacted by solid or liquid nonconductors. The charge that is transferred during contact electrification is stored on the surface of each object. Static electric generators, devices which produce very high voltage at very low current and used for classroom physics demonstrations, rely on this effect.

Presence of electric current does not detract from the electrostatic forces nor from the sparking, from the corona discharge, or other phenomena. Both phenomena can exist simultaneously in the same system.

### *Charge Neutralisation*

Natural electrostatic phenomena are most familiar as an occasional annoyance in seasons of low humidity, but can be destructive and harmful in some situations (e.g. electronics manufacturing). When working in direct contact with integrated circuit electronics (especially delicate MOSFETs), or in the presence of flammable gas, care must be taken to avoid accumulating and suddenly discharging a static charge.

### *Charge Induction*

Charge induction occurs when a negatively charged object repels electrons from the surface of a second object. This creates a region in the second object that is more positively charged. An attractive force is then exerted between the objects. For example, when a balloon is rubbed, the balloon will stick to the wall as an attractive force is exerted by two oppositely charged surfaces (the surface of the wall gains an electric charge due to charge induction, as the free electrons at the surface of the wall are repelled by the negative balloon, creating a positive wall surface, which is subsequently attracted to the surface of the balloon). You can explore the effect with a simulation of the balloon and static electricity.

### *"Static' Electricity*

Before the year 1832, when Michael Faraday published the results of his experiment on the identity of electricities, physicists thought "static electricity" was somehow different from other electrical charges.

Michael Faraday proved that the electricity induced from the magnet, voltaic electricity produced by a battery, and static electricity are all the same.

Static electricity is usually caused when certain materials are rubbed against each other, like wool on plastic or the soles of shoes on carpet. The process causes electrons to be pulled from the surface of one material and relocated on the surface of the other material.

A static shock occurs when the surface of the second material, negatively charged with electrons, touches a positively-charged conductor, or vice-versa.

Static electricity is commonly used in xerography, air filters, and some automotive paints. Static electricity is a build up of electric charges on two objects that have become separated from each other. Small electrical components can easily be damaged by static electricity. Component manufacturers use a number of antistatic devices to avoid this.

### Static Electricity and Chemical Industry

When different materials are brought together and then separated, an accumulation of electric charge can occur which leaves one material positively charged while the other becomes negatively charged. The mild shock that you receive when touching a grounded object after walking on carpet is an example of excess electrical charge accumulating in your body from frictional charging between your shoes and the carpet. The resulting charge build-up upon your body can generate a strong electrical discharge. Although experimenting with static electricity may be fun, similar sparks create severe hazards in those industries dealing with flammable substances, where a small electrical spark may ignite explosive mixtures with devastating consequences.

A similar charging mechanism can occur within low conductivity fluids flowing through pipelines—a process called flow electrification. Fluids which have low electrical conductivity (below 50 picosiemens per metre, where picosiemens per metre is a measure of electrical conductivity), are called accumulators. Fluids having conductivities above 50 pS/m are called non-accumulators. In non-accumulators, charges recombine as fast as they are separated and hence electrostatic charge generation is not significant. In the petrochemical industry, 50 pS/m is the recommended minimum value of electrical conductivity for adequate removal of charge from a fluid. An important concept for insulating fluids is the static relaxation time. This is similar to the time constant (tau) within an RC circuit. For insulating materials, it is the ratio of the static dielectric constant divided by the electrical conductivity of the material. For hydrocarbon fluids, this is sometimes approximated by dividing the number 18 by the electrical conductivity of the fluid. Thus a fluid that has an electrical conductivity of 1 pS/cm (100 pS/m) will have an estimated relaxation time of about 18 seconds. The excess charge within a fluid will be almost completely dissipated after 4 to 5 times the relaxation time, or 90 seconds for the fluid in the above example.

Charge generation increases at higher fluid velocities and larger pipe diametres, becoming quite significant in pipes 8 inches (200 mm) or larger. Static charge generation in these systems is best controlled by limiting fluid velocity. The British standard BS PD CLC/TR 50404:2003 (formerly BS-5958-Part 2) Code of Practice for Control of Undesirable Static Electricity prescribes velocity limits. Because of its large impact on dielectric constant, the recommended velocity for hydrocarbon fluids containing water should be limited to 1 m/s.

Bonding and earthing are the usual ways by which charge buildup can be prevented. For fluids with electrical conductivity below 10 pS/m, bonding and earthing are not adequate for charge dissipation, and anti-static additives may be required.

### Electrostatic Induction in Commercial Applications

The principle of electrostatic induction has been harnessed to beneficial effect in industry for many years, beginning with the introduction of electrostatic industrial painting systems for the economical and even application of enamel and polyurethane paints to consumer goods, including automobiles, bicycles, and other products.

# Chapter 4

## Molecular Self-assembly

Molecular self-assembly is the process by which molecules adopt a defined arrangement without guidance or management from an outside source. There are two types of self-assembly, intramolecular self-assembly and intermolecular self-assembly. Most often the term molecular self-assembly refers to intermolecular self-assembly, while the intramolecular analogue is more commonly called folding.

### Supramolecular Systems

Molecular self-assembly is a key concept in supramolecular chemistry since assembly of the molecules is directed through noncovalent interactions (e.g., hydrogen bonding, metal coordination, hydrophobic forces, van der Waals forces, $\pi$-$\pi$ interactions, and/or electrostatic) as well as electromagnetic interactions. Common examples include the formation of micelles, vesicles, liquid crystal phases, and Langmuir monolayers by surfactant molecules. Further examples of supramolecular assemblies demonstrate that a variety of different shapes and sizes can be obtained using molecular self-assembly.

Molecular self-assembly has allowed the construction of challenging molecular topologies. An example are Borromean rings, interlocking rings wherein removal of one ring unlocks each of the other rings. DNA has been used to prepare a molecular analogue of Borromean rings. More recently, a similar structure has been prepared using non-biological building blocks.

### Biological Systems

Molecular self-assembly is crucial to the function of cells. It is exhibited in the self-assembly of lipids to form the membrane, the formation of double helical DNA through hydrogen bonding of the individual strands, and the assembly of proteins to form quaternary structures. Molecular self-assembly of incorrectly folded proteins into insoluble amyloid fibers is responsible for infectious prion-related neurodegenerative diseases.

## Nanotechnology

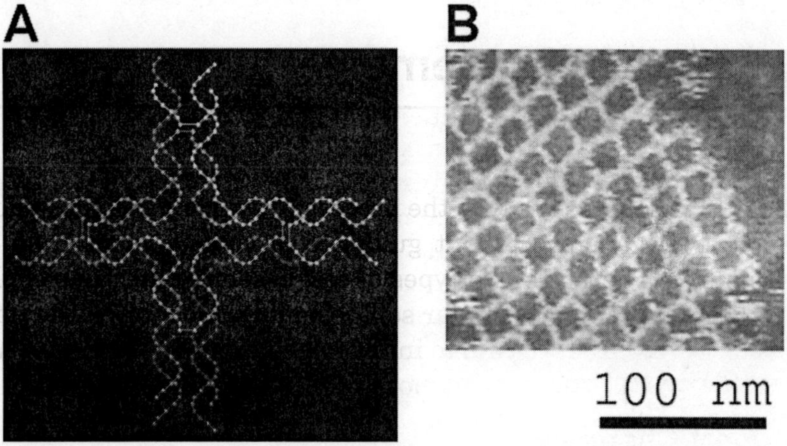


**Figure 1:** The DNA structure at left will self-assemble into the structure visualised by atomic force microscopy at right. Image from Strong.

Molecular self-assembly is an important aspect of bottom-up approaches to nanotechnology. Using molecular self-assembly the final (desired) structure is programmed in the shape and functional groups of the molecules. Self-assembly is referred to as a 'bottom-up' manufacturing technique in contrast to a 'top-down' technique such as lithography where the desired final structure is carved from a larger block of matter. In the speculative vision of molecular nanotechnology, microchips of the future might be made by molecular self-assembly. An advantage to constructing nanostructure using molecular self-assembly for biological materials is that they will degrade back into individual molecules that can be broken down by the body.

### *DNA Nanotechnology*

DNA nanotechnology is an area of current research that uses the bottom-up, self-assembly approach for nanotechnological goals. DNA nanotechnology uses the unique molecular recognition properties of DNA and other nucleic acids to create self-assembling branched DNA complexes with useful properties. DNA is thus used as a structural material rather than as a carrier of biological information, to make structures such as two-dimensional periodic lattices (both tile-based as well as using the "DNA origami" method) and three-dimensional structures in the shapes of polyhedra. These DNA structures have also been used to template the assembly of other molecules such as gold nanoparticles and streptavidin proteins.

### *Two-dimensional Monolayers*

The spontaneous assembly of a single layer of molecules (i.e. monolayer thick) at interfaces is usually referred to as two-dimensional self-assembly. Early direct proofs showing that molecules can assembly into higher-order architectures at solid interfaces came with the development of scanning tunnelling microscopy and shortly thereafter. Eventually two strategies became popular for the self-assembly of 2D architectures, namely self-assembly following ultra-high-vacuum deposition and annealing and self-assembly at the solid-liquid interface. The design of molecules and conditions leading to the formation of highly-crystalline architectures is considered today a form of 2D crystal engineering at then anoscopic scale.

## Host-guest Chemistry

In supramolecular chemistry, host-guest chemistry describes complexes that are composed of two or more molecules or ions that are held together in unique structural relationships by forces other than those of full covalent bonds. Host-guest chemistry encompasses the idea of molecular recognition and interactions through noncovalent bonding. Noncovalent bonding is critical in maintaining the three-dimensional structure of large molecules, such as proteins and is involved in many biological processes in which large molecules bind specifically but transiently to one another. There are four commonly mentioned types of non-covalent interactions: hydrogen bonds, ionic bonds, van der Waals forces, and hydrophobic interactions.

### *Common Host Molecules*

Commons host molecules are cyclodextrins, calixarenes, cucurbiturils, porphyrins , metallacrowns, crown ethers, zeolites, cyclotriveratrylenes, cryptophanes and carcerands.

Host-guest chemistry is observed in inclusion compounds, Intercalation compounds, clathrates and molecular tweezers.

### *Thermodynamic Principles of Host-Guest Interactions*

There is an equilibrium between the unbound state, in which the host and the guest are separate from each other, and the bound state, in which there is a structurally defined host-guest complex:

$$H + G \rightleftharpoons HG$$

H ="host" , G ="guest" , HG ="host-guest complex"

The "host" component can be considered the larger molecule, and it encompasses the smaller, "guest", molecule. In biological systems, the analogous terms of host and guest are commonly referred to as enzyme and substrate respectively.

The thermodynamic benefits of host guest chemistry are derived from the idea that there is a lower overall Gibbs free energy due to the interaction between host and guest molecules. Chemists are exhaustively trying to measure the energy and thermodynamic properties of these non-covalent interactions found throughout supramolecular chemistry; and by doing so hope to gain further insight into the combinatorial outcome of these many, small, non-covalent forces that are used to generate an overall effect on the supramolecular structure.

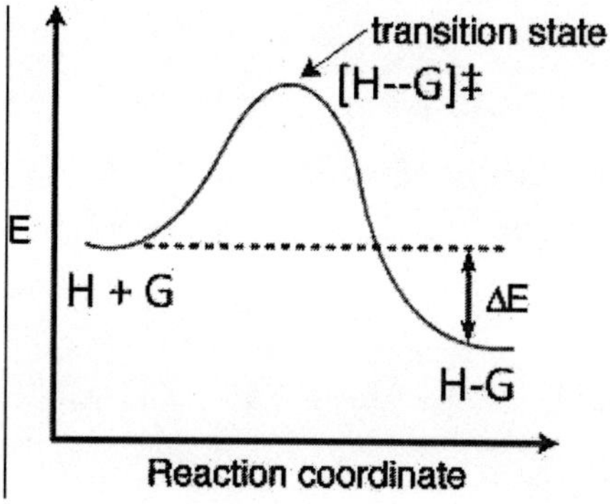


**Figure 2:** Free Energy Diagram of Host-Guest Complex Formation. X axis = Reaction Coordinate. Y axis = Free Energy. One can see that the host-guest complex is at a lower overall energy that the individual molecules

In order to rationally and confidently design synthetic systems that perform specific functions and tasks, it is very important to understand the thermodynamics of binding between host and guest. Chemists are focusing on the energy exchange of different binding interactions and trying to develop scientific experiments to quantify the fundamental origins of these non-covalent interactions by utilising various techniques such as NMR spectroscopy, Raman spectroscopy, isothermal titration calorimetry, surface tension, and UV-Vis Spectroscopy. The experimental data is quantified and explained through analysis of binding constants $K_a$, Gibbs free energy $\Delta G_o$, Enthalpy $\Delta H_o$, and entropy $\Delta S_o$.

### Association and Dissociation Constants

$$K_a = \frac{[HG]_{eq}}{[H]_{eq}[G]_{eq}}$$

The association constant, $K_a$, is equal to the concentration of the Host-Guest complex divided by the product of the concentrations of the individual Host and Guest molecules when the system is in equilibrium.

The equilibrium that has been established between the Host-Guest complex and free molecules can also be defined by a dissociation constant, $K_d$.

$$K_d = \frac{[H]_{eq}[G]_{eq}}{[HG]_{eq}} = \frac{1}{K_a}$$

A Large $K_a$ value can be equated to a small $K_d$, both values indicate a strong complexation between host and guest molecules to form the host-guest complex.

### Gibbs Free Energy dependency on K

The Change in Gibbs Free Energy, ΔG, is a function of the equilibrium constant, $K_a$

$\Delta G = RT\ln K_a$

One can see that by knowing the association constant, you can solve for the Gibbs free energy of the reaction.

$$\Delta G = \Delta H - T\Delta S,$$

One can also can see the dependency of Gibbs Free energy on the macroscopic terms of Enthalpy, ΔH and Entropy, ΔS.

### Experimental Determination of [HG] using Kinetics

In order to determine the concentration of the *HG* complex, one needs to write the equilibrium equation in terms that are able to be measured in the laboratory. Experimentally, one can determine the concentration of the host-guest complex, $[HG]$, as a function of the initial concentration of the guest molecules, $[G]_o$.

Their occurs a problem because our association constant, which one wants to relate to the Gibbs Free Energy of the system, is a function of the concentration of host, $[H]_{eq}$ and guest, $[G]_{eq}$, molecules at equilibrium. Therefore, in order to be able to successfully determine the $[HG]$ as a function of $[G]_o$, one must make some assumptions.

These assumptions are necessary because the scientist only known the initial concentrations of host and guest, $[H]_o$ and $[G]_o$

In order to overcome this obstacle, one can first make the alternative equilibrium approximation which states that the initial concentration of host molecules, $[H]_o$ is equivalent to the concentration of both $[HG]_{eq}$ and $[H]_{eq}$; this is because sum of all $[H]$ is only found in $[HG]_{eq}$ and $[H]_{eq}$

$[H]_o$ in terms of $[HG]_{eq}$ and $[H]_{eq}$:

$[H]_o = [HG]_{eq} + [H]_{eq}$

Solving for $[H]$ yields:

$[H]_{eq} = [H]_o$ " $[HG]_{eq}$

This can be substituted into the original equilibrium equation

$$K_a = \frac{[HG]_{eq}}{[H_{eq}][G_{eq}]}$$

$$K_a = \frac{[HG]_{eq}}{[G]_{eq}([H]_o - [HG]_{eq})}$$

In solving for $[HG]_{eq}$, one can determine that:

$$[HG]_{eq} = \frac{K_a[G]_{eq}[H]_o}{1 + K_a[G]_{eq}}$$

At this point, we have gotten the concentration of $[HG]_{eq}$ complex in terms of the initial concentration of host molecules, $[H]_o$

## Systems Biology

Systems biology is a term used to describe a number of trends in bioscience research, and a movement which draws on those trends. Proponents describe systems biology as a biology-based inter-disciplinary study field that focuses on complex interactions in biological systems, claiming that it uses a new perspective (holism instead of reduction).

Particularly from year 2000 onwards, the term is used widely in the biosciences, and in a variety of contexts. An often stated ambition of systems biology is the modeling and discovery of emergent properties, properties of a system whose theoretical description is only possible using techniques which fall under the remit of systems biology. These typically involve cell signalling networks, via long-range allostery.

### *Overview*

Systems biology can be considered from a number of different aspects:

- As a field of study, particularly, the study of the interactions between the components of *biological systems*, and how these interactions give rise to the function and behaviour of that system (for example, the enzymes and metabolites in a metabolic pathway).
- As a paradigm, usually defined in antithesis to the so-called reductionist paradigm (biological organisation), although fully consistent with the scientific method. The distinction between the two paradigms is referred to in these quotations:

  "The reductionist approach has successfully identified most of the components and many of the interactions but, unfortunately, offers no convincing concepts or methods to understand how system properties emerge...the pluralism of causes and effects in biological networks is better addressed by observing, through quantitative measures, multiple components simultaneously and by rigorous data integration with mathematical models" *Sauer et al*

  "Systems biology...is about putting together rather than taking apart, integration rather than reduction. It requires that we develop ways of thinking about integration that are as rigorous as our reductionist programmes, but different....It means changing our philosophy, in the full sense of the term" *Denis Noble*
- As a series of operational protocols used for performing research, namely a cycle composed of theory, analytic or computational modelling to propose specific testable hypotheses about a biological system, experimental validation, and then using the newly acquired quantitative description of cells or cell processes to refine the computational model or theory. Since the objective is a model of the interactions in a system, the experimental techniques that most suit systems biology are those that are system-wide and attempt to be as complete as possible. Therefore, transcriptomics, metabolomics, proteomics and high-throughput techniques are used to collect quantitative data for the construction and validation of models.

- As the application of dynamical systems theory to molecular biology.
- As a socioscientific phenomenon defined by the strategy of pursuing integration of complex data about the interactions in biological systems from diverse experimental sources using interdisciplinary tools and personnel.

This variety of viewpoints is illustrative of the fact that systems biology refers to a cluster of peripherally overlapping concepts rather than a single well-delineated field. However the term has widespread currency and popularity as of 2007, with chairs and institutes of systems biology proliferating worldwide.

### *History*

Systems biology finds its roots in:

- the quantitative modeling of enzyme kinetics, a discipline that flourished between 1900 and 1970,
- the mathematical modeling of population growth,
- the simulations developed to study neurophysiology, and
- control theory and cybernetics.

One of the theorists who can be seen as one of the precursors of systems biology is Ludwig von Bertalanffy with his general systems theory. One of the first numerical simulations in biology was published in 1952 by the British neurophysiologists and Nobel prize winners Alan Lloyd Hodgkin and Andrew Fielding Huxley, who constructed a mathematical model that explained the action potential propagating along the axon of a neuronal cell. Their model described a cellular function emerging from the interaction between two different molecular components, a potassium and a sodium channel, and can therefore be seen as the beginning of computational systems biology. In 1960, Denis Noble developed the first computer model of the heart pacemaker.

The formal study of systems biology, as a distinct discipline, was launched by systems theorist Mihajlo Mesarovic in 1966 with an international symposium at the Case Institute of Technology in Cleveland, Ohio entitled "Systems Theory and Biology".

The 1960s and 1970s saw the development of several approaches to study complex molecular systems, such as the Metabolic Control Analysis and the biochemical systems theory. The successes of molecular biology throughout the 1980s, coupled with a skepticism toward theoretical biology, that then promised more than it achieved, caused the quantitative modelling of biological processes to become a somewhat minor field.

However the birth of functional genomics in the 1990s meant that large quantities of high quality data became available, while the computing power exploded, making more realistic models possible. In 1997, the group of Masaru Tomitapublished the first quantitative model of the metabolism of a whole (hypothetical) cell.

Around the year 2000, after Institutes of Systems Biology were established in Seattle and Tokyo, systems biology emerged as a movement in its own right, spurred on by the completion of various genome projects, the large increase in data from the omics (e.g. genomics and proteomics) and the accompanying advances in high-throughput experiments and bioinformatics.

Since then, various research institutes dedicated to systems biology have been developed. For example, the NIGMS of NIH established a project grant that is currently supporting over ten Systems Biology Centres in the United States. As of summer 2006, due to a shortage of people in systems biology several doctoral training programs in systems biology have been established in many parts of the world. In that same year, the National Science Foundation (NSF) put forward a grand challenge for systems biology in the 21st century to build a mathematical model of the whole cell. In 2011, V. A. Shiva Ayyadurai and C. Forbes Dewey, Jr. of Department of Biological Engineering at the Massachusetts Institute of Technology created CytoSolve, a method to model the whole cell by dynamically integrating multiple molecular pathway models.

### Associated disciplines

According to the interpretation of Systems Biology as the ability to obtain, integrate and analyse complex data sets from multiple experimental sources using interdisciplinary tools, some typical technology platforms are:

- *Phenomics:* Organismal variation in phenotype as it changes during its life span.
- *Genomics:* Organismal deoxyribonucleic acid (DNA) sequence, including intra-organisamal cell specific variation. (i.e. Telomere length variation etc.).

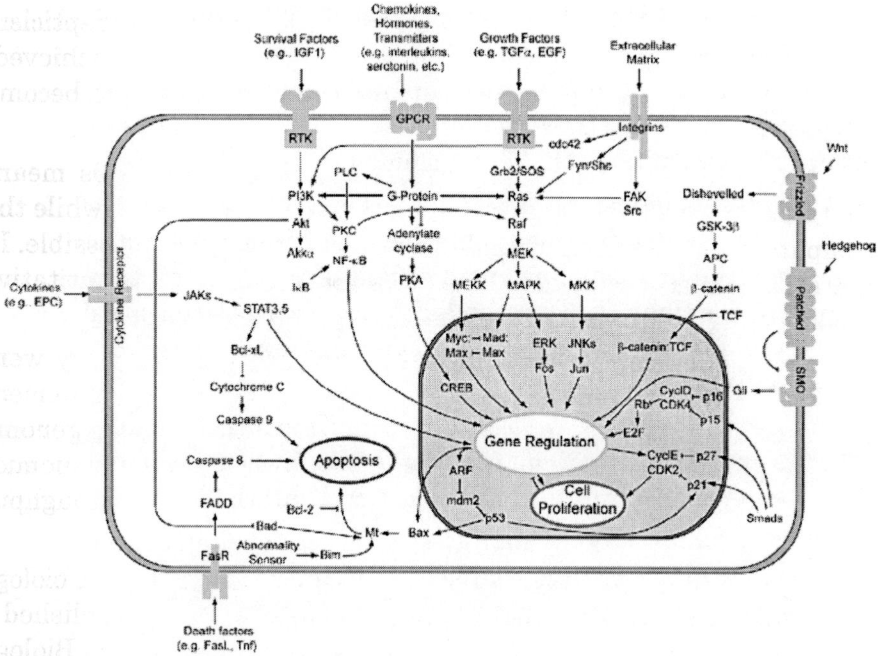


**Figure 3:** Overview of signal transduction pathways

- *Epigenomics / Epigenetics:* Organismal and corresponding cell specific transcriptomic regulating factors not empirically coded in the genomic sequence. (i.e. DNA methylation, Histone Acetelation etc.).
- *Transcriptomics:* Organismal, tissue or whole cell gene expression measurements by DNA microarrays or serial analysis of gene expression
- *Interferomics:* Organismal, tissue, or cell level transcript correcting factors (i.e. RNA interference)
- *Translatomics / Proteomics:* Organismal, tissue, or cell level measurements of proteins and peptides via two-dimensional gel electrophoresis, mass spectrometry or multidimensional protein identification techniques (advanced HPLC systems coupled with mass spectrometry). Sub disciplines include phosphoproteomics, glycoproteomics and other methods to detect chemically modified proteins.
- *Metabolomics:* Organismal, tissue, or cell level measurements of all small-molecules known as metabolites.
- *Glycomics:* Organismal, tissue, or cell level measurements of carbohydrates.

- *Lipidomics:* Organismal, tissue, or cell level measurements of lipids.

In addition to the identification and quantification of the above given molecules further techniques analyse the dynamics and interactions within a cell. This includes:

- *Interactomics:* Organismal, tissue, or cell level study of interactions between molecules. Currently the authoritative molecular discipline in this field of study is protein-protein interactions (PPI), although the working definition does not pre-clude inclusion of other molecular disciplines such as those defined here.
- *NeuroElectroDynamics:* Organismal, brain computing function as a dynamic system, underlying biophysical mechanisms and emerging computation by electrical interactions.
- *Fluxomics:* Organismal, tissue, or cell level measurements of molecular dynamic changes over time.
- Biomics: systems analysis of the biome.

The investigations are frequently combined with large-scale perturbation methods, including gene-based (RNAi, mis-expression of wild type and mutant genes) and chemical approaches using small molecule libraries. Robots and automated sensors enable such large-scale experimentation and data acquisition.

These technologies are still emerging and many face problems that the larger the quantity of data produced, the lower the quality. A wide variety of quantitative scientists (computational biologists, statisticians, mathematicians, computer scientists, engineers, and physicists) are working to improve the quality of these approaches and to create, refine, and retest the models to accurately reflect observations.

The systems biology approach often involves the development of mechanistic models, such as the reconstruction of dynamic systems from the quantitative properties of their elementary building blocks. For instance, a cellular network can be modelled mathematically using methods coming from chemical kinetics and control theory. Due to the large number of parameters, variables and constraints in cellular networks, numerical and computational techniques are often used.

Other aspects of computer science and informatics are also used in systems biology. These include:

- New forms of computational model, such as the use of process calculi to model biological processes (notable approaches include stochastic ð-calculus, BioAmbients, Beta Binders, BioPEPA and Brane calculus) and constraint-based modeling.
- Integration of information from the literature, using techniques of information extraction and text mining.
- Development of online databases and repositories for sharing data and models, approaches to database integration and software interoperability via loose coupling of software, websites and databases, or commercial suits.
- Development of syntactically and semantically sound ways of representing biological models

## Noncovalent Bonding

A noncovalent bond is a type of chemical bond, typically between macromolecules, that does not involve the sharing of pairs of electrons, but rather involves more dispersed variations of electromagnetic interactions. The noncovalent bond is the dominant type of bond between supermolecules in supermolecular chemistry. Noncovalent bonds are critical in maintaining the three-dimensional structure of large molecules, such as proteins and nucleic acids, and are involved in many biological processes in which large molecules bind specifically but transiently to one another.

The energy released in the formation of noncovalent bonds is on the order of 1-5 kcal per mol. There are four commonly mentioned types of non-covalent interactions: hydrogen bonds, ionic bonds, van der Waals forces, and hydrophobic interactions. The noncovalent interactions hold together the two strands of DNA in the double helix, stabilise secondary and tertiary structures of proteins, and enable enzyme-substrate binding and antibody-antigen association.

### Overview

In general, *noncovalent bonding* refers to attractive intermolecular forces that are not covalent in nature. Noncovalent forces are dominant in supramolecular chemistry. These noncovalent interactions may include ionic bonds, hydrophobic interactions, hydrogen bonds and van der Waals forces (dispersion attractions, dipole-dipole and dipole-induced dipole interactions). Noncovalent bonds are weak by nature and must therefore work together to have a significant effect. In addition, the combined bond strength is greater than the sum of the individual bonds. This is because the free energy of multiple bonds between two molecules is greater than the sum of the enthalpies of each bond due to [[entropy | entropic].

### Examples

***Protein Structure:*** Intramolecular noncovalent interactions are largely responsible for the secondary and tertiary structure of proteins and therefore the protein's function in the mechanisms of life. Intermolecular noncovalent interactions are responsible for protein complexes (quaternary structure) where two or more proteins function in a coherent mechanism.

### Pharmaceuticals

Most drugs work by noncovalently interacting with biomolecules such as proteins or RNA. Relatively few drugs actually form covalent bonds with the biomolecules they interact with; instead, they interfere with or activate some biological mechanism through noncovalently interacting in very specific locations on specific biomolecules which present the perfect combination of noncovalent binding partners in just the right geometry.

# Chapter 5

# Protein Structure

Proteins are an important class of biological macromolecules present in all organisms. Proteins are polymers of amino acids. Classified by their physical size, proteins are nanoparticles (definition: 1–100 nm). Each protein polymer – also known as a polypeptide – consists of a sequence formed from 20 possible L-α-amino acids, also referred to as residues. For chains under 40 residues the term peptide is frequently used instead of protein. To be able to perform their biological function, proteins fold into one or more specific spatial conformations, driven by a number of non-covalent interactions such as hydrogen bonding, ionic interactions, Van Der Waals forces, and hydrophobic packing. To understand the functions of proteins at a molecular level, it is often necessary to determine their three-dimensional structure. This is the topic of the scientific field of structural biology, which employs techniques such as X-ray crystallography, NMR spectroscopy, anddual polarisation interferometry to determine the structure of proteins.

Protein structures range in size from tens to several thousand residues Very large aggregates can be formed from protein subunits: for example, many thousand actin molecules assemble into a microfilament. A protein may undergo reversible structural changes in performing its biological function. The alternative structures of the same protein are referred to as different conformations, and transitions between them are called conformational changes.

## Protein Covalent Structure and Stereochemistry

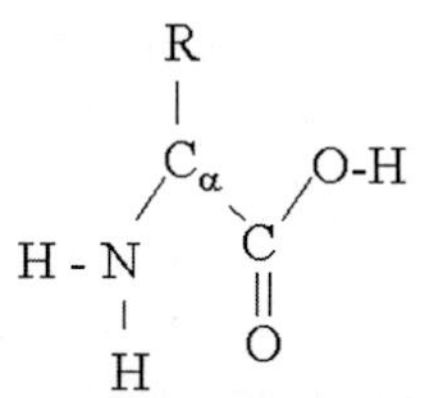


**Figure 1:** An α-amino acid. The $C_{\alpha}H$ atom is omitted in the diagram.

Protein amino acids are combined into a single polypeptide chain in a condensation reaction. This reaction is catalysed by the ribosome in a process known as translation.

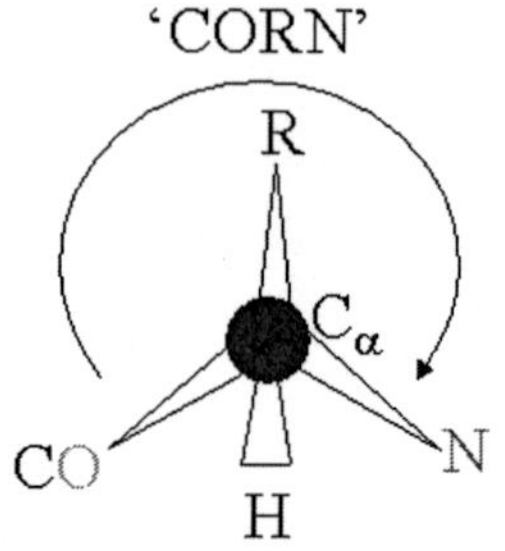


**Figure 2:** CO-R-N rule

The 20 naturally occurring amino acids have different physical and chemical properties, including their electrostatic charge, $pK_a$, hydrophobicity, size and specific functional groups. These properties play a major role in molding protein structure.

## The Peptide Bond

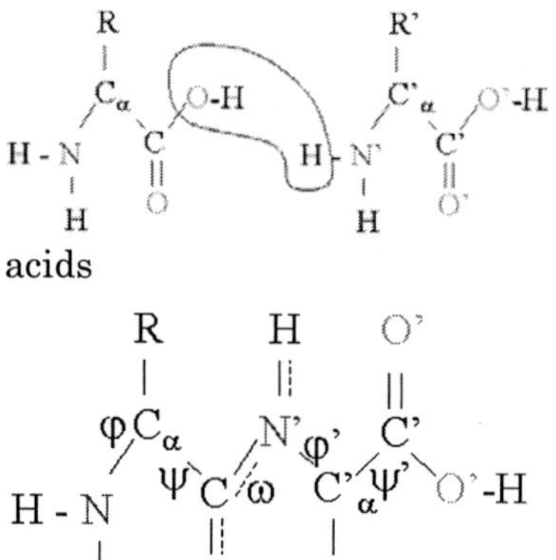

**Figure 3:** Two amino acids

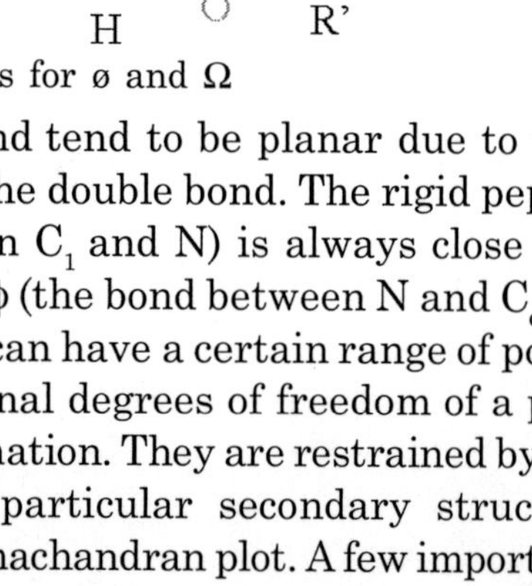

**Figure 4:** Bond angles for ø and Ω

The peptide bond tend to be planar due to the delocalisation of the electrons from the double bond. The rigid peptide dihedral angle, Ω (the bond between $C_1$ and N) is always close to 180 degrees. The dihedral angles phi ϕ (the bond between N and $C_\alpha$) and psi ø (the bond between $C_\alpha$ and $C_1$) can have a certain range of possible values. These angles are the internal degrees of freedom of a protein, they control the protein's conformation. They are restrained by geometry to allowed ranges typical for particular secondary structure elements, and represented in a Ramachandran plot. A few important bond lengths are given in the table below.

| *Peptide bond* | *Average length* | *Single bond* | *Average length* | *Hydrogen bond* | *Average (±30)* |
|---|---|---|---|---|---|
| $C_\alpha$–C | 153 pm | C–C | 154 pm | O–H — O–H | 280 pm |
| C–N | 133 pm | C–N | 148 pm | N–H — O=C | 290 pm |
| N–$C_\alpha$ | 146 pm | C–O | 143 pm | O–H — O=C | 280 pm |

### *Side-chain Conformation*

The atoms along the side chain are named with Greek letters in Greek alphabetical order: α, β, γ, Δ, °, and so on. $C_\alpha$ refers to the carbon atom of the backbone closest to the carbonyl group of that amino acid, $C_\beta$ the second closest and so on. The dihedral angles around the bonds between these atoms are named ÷1, ÷2, ÷3, etc. The dihedral angle of the first movable atom of the side chain, γ, defined as N-Cα-Cβ-*X*γ, is named ÷1. Side chains tend to adopt different staggered conformations called *gauche(-), trans*, and *gauche(+)*, which corresponds to rotation angles of 60°, 180°, and -60°, respectively, around the sp3-sp3 bonds.

The diversity of side-chain conformations is often expressed in rotamer libraries. A rotamer library is a collection of rotamers for each residue type. Side-chain dihedral angles are not evenly distributed, but for most side chain types, the ÷ angles occur in tight clusters around certain values. Rotamer libraries therefore are usually derived from statistical analysis of side-chain conformations in known structures of proteins by clustering observed conformations or by dividing dihedral angle space into bins, and determining an average conformation in each bin.

## Levels of Protein Structure

There are four distinct levels of protein structure.

### *Primary Structure*

The primary structure refers to amino acid sequence of the polypeptide chain. The primary structure is held together by covalent or peptide bonds, which are made during the process of protein biosynthesis or translation. The two ends of the polypeptide chain are referred to as the carboxyl terminus (C-terminus) and the amino terminus (N-terminus) based on the nature of the free group on each extremity. Counting of residues always starts at the N-terminal end ($NH_2$-group), which is the end where the amino group is not involved in a peptide bond.

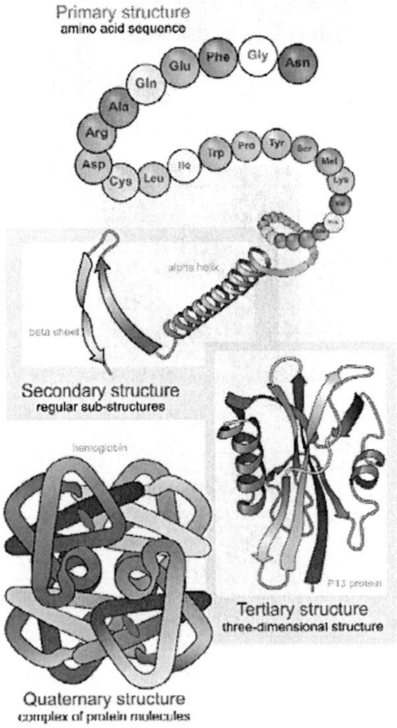


**Figure 5:** Protein structure, from primary to quaternary structure.

The primary structure of a protein is determined by the gene corresponding to the protein. A specific sequence of nucleotides in DNA is transcribed into mRNA, which is read by the ribosome in a process called translation. The sequence of a protein is unique to that protein, and defines the structure and function of the protein. The sequence of a protein can be determined by methods such as Edman degradation or tandem mass spectrometry. Often however, it is read directly from the sequence of the gene using the genetic code. Post-translational modifications such as disulfide formation, phosphorylations and glycosylations are usually also considered a part of the primary structure, and cannot be read from the gene.

## Amino Acid Residues

Each α-amino acid consists of a backbone part that is present in all the amino acid types, and a side chain that is unique to each type of residue. An exception from this rule is proline. Because the carbon atom is bound to four different groups it is chiral, however only one of the isomers occur in biological proteins. Glycine however, is not chiral since its side chain is a hydrogen atom. A simple mnemonic for correct L-form is "CORN": when the $C_\alpha$ atom is viewed with the H in front, the residues read "CO-R-N" in a clockwise direction.

## Secondary Structure

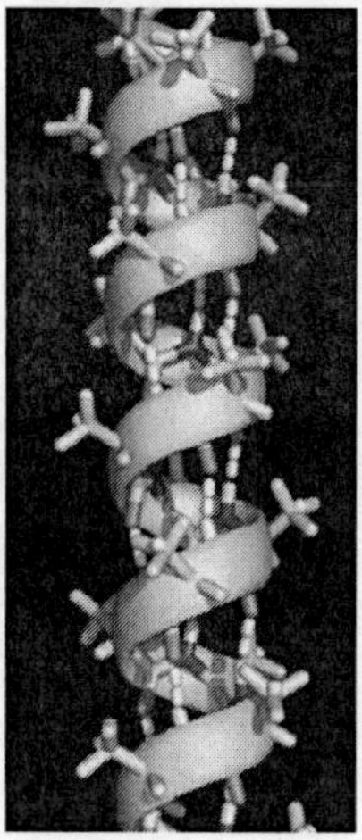

**Figure 6:** An alpha-helix with hydrogen bonds (yellow dots)

Secondary structure refers to highly regular local sub-structures. Two main types of secondary structure, the alpha helix and the beta strand, were suggested in 1951 by Linus Pauling and coworkers. These secondary structures are defined by patterns of hydrogen bonds between the main-chain peptide groups. They have a regular geometry, being constrained to specific values of the dihedral angles ø and ϕ on the Ramachandran plot. Both the alpha helix and the beta-sheet represent a way of saturating all the hydrogen bond donors and acceptors in the peptide backbone. Some parts of the protein are ordered but do not form any regular structures. They should not be confused with random coil, an unfolded polypeptide chain lacking any fixed three-dimensional structure. Several sequential secondary structures may form a "supersecondary unit".

## Tertiary Structure

Tertiary structure refers to three-dimensional structure of a single protein molecule. The alpha-helices and beta-sheets are folded into a compact globule. The folding is driven by the *non-specific* hydrophobic interactions (the burial of hydrophobic residues from water), but the structure is stable only when the parts of a protein domain are locked into place by *specific* tertiary interactions, such as salt bridges, hydrogen bonds, and the tight packing of side chains and disulfide bonds. The disulfide bonds are extremely rare in cytosolic proteins, since the cytosol is generally a reducing environment.

## Quaternary Structure

Quaternary structure is a larger assembly of several protein molecules or polypeptide chains, usually called subunits in this context. The quaternary structure is stabilised by the same non-covalent interactions and disulfide bonds as the tertiary structure. Complexes of two or more polypeptides (i.e. multiple subunits) are called multimers. Specifically it would be called a dimer if it contains two subunits, a trimer if it contains three subunits, and a tetramer if it contains four subunits. The subunits are frequently related to one another by symmetry operations, such as a 2-fold axis in a dimer. Multimers made up of identical subunits are referred to with a prefix of "homo-" (e.g. a homotetramer) and those made up of different subunits are referred to with a prefix of "hetero-" (e.g. a heterotetramer, such as the two alpha and two beta chains of hemoglobin).

## Domains, Motifs, and Folds in Protein Structure

Protein are frequently described as consisting from several structural units.

- A structural domain is an element of the protein's overall structure that is self-stabilising and often folds independently of the rest of the protein chain. Many domains are not unique to the protein products of one gene or one gene family but instead appear in a variety of proteins. Domains often are named and singled out because they figure prominently in the biological function of the protein they belong to; for example, the "calcium-binding domain of calmodulin". Because they are independently stable, domains can be "swapped" by genetic engineering between one protein and another to make chimeras.
- The structural and sequence motifs refer to short segments of protein three-dimensional structure or amino acid sequence that were found in a large number of different proteins.
- The supersecondary structure refers to a specific combination of secondary structure elements, such as beta-alpha-beta units

or helix-turn-helix motif. Some of them may be also referred to as structural motifs.

- Protein fold refers to the general protein architecture, like helix bundle, beta-barrel, Rossman fold or different "folds" provided in the Structural Classification of Proteins database.

Despite the fact that there are about 100,000 different proteins expressed in eukaryotic systems, there are many fewer different domains, structural motifs and folds. This is partly a consequence of evolution, since genes or parts of genes can be doubled or moved around within the genome. This means that, for example, a protein domain might be moved from one protein to another thus giving the protein a new function. Because of these mechanisms, pathways and mechanisms tend to be reused in several different proteins.

### Protein Folding

An unfolded polypeptide folds into its characteristic three-dimensional structure from random coil.

### Protein Structure Determination

Around 90% of the protein structures available in the Protein Data Bank have been determined by X-ray crystallography. This method allows one to measure the 3D density distribution of electrons in the protein (in the crystallised state) and thereby infer the 3D coordinates of all the atoms to be determined to a certain resolution. Roughly 9% of the known protein structures have been obtained by Nuclear Magnetic Resonance techniques.

The secondary structure composition can be determined via circular dichroism or dual polarisation interferometry. Cryo-electron microscopy has recently become a means of determining protein structures to high resolution (less than 5 angstroms or 0.5 nanometre) and is anticipated to increase in power as a tool for high resolution work in the next decade. This technique is still a valuable resource for researchers working with very large protein complexes such as virus coat proteins and amyloid fibers.

## Structure Classification

Protein structures can be grouped based on their similarity or a common evolutionary origin. SCOP and CATH databases provide two different structural classifications of proteins.

### *Computational Prediction of Protein Structure*

The generation of a protein sequence is much easier than the determination of a protein structure. However, the structure of a protein gives much more insight in the function of the protein than its sequence. Therefore, a number of methods for the computational prediction of protein structure from its sequence have been developed. *Ab initio* prediction methods use just the sequence of the protein. Threading and Homology Modeling methods can build a 3D model for a protein of unknown structure from experimental structures of evolutionary related proteins.

## Nucleic Acid Double Helix

In molecular biology, the term double helix refers to the structure formed by double-stranded molecules of nucleic acids such as DNA and RNA. The double helical structure of a nucleic acid complex arises as a consequence of its secondary structure, and is a fundamental component in determining its tertiary structure. The term entered popular culture with the publication in 1968 of *The Double Helix: A Personal Account of the Discovery of the Structure of DNA*, by James Watson.

The DNA double helix is a spiral polymer of nucleic acids, held together by nucleotides which base pair together. In B-DNA, the most common double helical structure, the double helix is right-handed with about 10–10.5 nucleotides per turn. The double helix structure of DNA contains a *major groove* and *minor groove*, the major groove being wider than the minor groove. Given the difference in widths of the major groove and minor groove, many proteins which bind to DNA do so through the wider major groove.

### *History*

The double-helix model of DNA structure was first published in the journal *Nature* by James D. Watson and Francis Crick in 1953, (X,Y,Z coordinates in 1954) based upon the crucial X-ray diffraction image of DNA labelled as “Photo 51”, from Rosalind Franklin in 1952, followed by her more clarified DNA image with Raymond Gosling, Maurice Wilkins, Alexander Stokes, and Herbert Wilson, as well as base-pairing chemical and biochemical information by Erwin Chargaff. The previous model was triple-stranded DNA.

The realisation that the structure of DNA is that of a double-helix elucidated the mechanism of base pairing by which genetic information is stored and copied in living organisms, and is widely considered one of the most important scientific discoveries of the 20th century. Crick, Wilkins, and Watson each received one third of the 1962 Nobel Prize in Physiology or Medicine for their contributions to the discovery. (Franklin, whose breakthrough X-ray diffraction data was used to formulate the DNA structure, died in 1958, and thus was ineligible to be nominated for a Nobel Prize.)

### Nucleic Acid Hybridisation

Hybridisation is the process of complementary base pairs binding to form a double helix. Melting is the process by which the interactions between the strands of the double helix are broken, separating the two nucleic acid strands. These bonds are weak, easily separated by gentle heating, enzymes, or physical force. Melting occurs preferentially at certain points in the nucleic acid. T and A rich sequences are more easily melted than C and G rich regions. Particular base steps are also susceptible to DNA melting, particularly T A and T G base steps. These mechanical features are reflected by the use of sequences such as TATAA at the start of many genes to assist RNA polymerase in melting the DNA for transcription.

Strand separation by gentle heating, as used in PCR, is simple providing the molecules have fewer than about 10,000 base pairs (10 kilobase pairs, or 10 kbp). The intertwining of the DNA strands makes long segments difficult to separate. The cell avoids this problem by allowing its DNA-melting enzymes (helicases) to work concurrently with topoisomerases, which can chemically cleave the phosphate backbone of one of the strands so that it can swivel around the other. Helicases unwind the strands to facilitate the advance of sequence-reading enzymes such as DNA polymerase.

### Base Pair Geometry

The geometry of a base, or base pair step can be characterised by 6 coordinates: Shift, slide, rise, tilt, roll, and twist. These values precisely define the location and orientation in space of every base or base pair in a nucleic acid molecule relative to its predecessor along the axis of the helix. Together, they characterise the helical structure of the molecule. In regions of DNA or RNA where the "normal" structure is disrupted, the change in these values can be used to describe such disruption. For each base pair, considered relative to its predecessor, there are the following base pair geometries to consider:

- Shear
- Stretch
- Stagger
- Buckle
- Propeller twist: rotation of one base with respect to the other in the same base pair.
- Opening
- Shift: displacement along an axis in the base-pair plane perpendicular to the first, directed from the minor to the major groove.
- Slide: displacement along an axis in the plane of the base pair directed from one strand to the other.
- Rise: displacement along the helix axis.
- Tilt: rotation around this axis.
- vRoll: rotation around this axis.
- Twist: rotation around the helix axis.
- vx-displacement
- y-displacement
- inclination
- tip
- pitch: the number of base pairs per complete turn of the helix.

Rise and twist determine the handedness and pitch of the helix. The other coordinates, by contrast, can be zero. Slide and shift are typically small in B-DNA, but are substantial in A- and Z-DNA. Roll and tilt make successive base pairs less parallel, and are typically small. A diagram of these coordinates can be found in 3DNA.

"Tilt" has often been used differently in the scientific literature, referring to the deviation of the first, inter-strand base-pair axis from perpendicularity to the helix axis. This corresponds to slide between a succession of base pairs, and in helix-based coordinates is properly termed "inclination".

### Helix Geometries

At least three DNA conformations are believed to be found in nature, A-DNA, B-DNA, and Z-DNA. The "B" form described by James

D. Watson and Francis Crick is believed to predominate in cells. It is 23.7 E wide and extends 34 E per 10 bp of sequence.

The double helix makes one complete turn about its axis every 10.4-10.5 base pairs in solution. This frequency of twist (known as the helical *pitch*) depends largely on stacking forces that each base exerts on its neighbours in the chain.

A-DNA and Z-DNA differ significantly in their geometry and dimensions to B-DNA, although still form helical structures. The A form appears likely to occur only in dehydrated samples of DNA, such as those used in crystallographic experiments, and possibly in hybrid pairings of DNA and RNA strands.

Segments of DNA that cells have been methylated for regulatory purposes may adopt the Z geometry, in which the strands turn about the helical axis the opposite way to A-DNA and B-DNA. There is also evidence of protein-DNA complexes forming Z-DNA structures.

Other conformations are possible; A-DNA, B-DNA, C-DNA, E-DNA, L-DNA (the enantiomeric form of D-DNA), P-DNA, S-DNA, Z-DNA, etc. have been described so far. In fact, only the letters F, Q, U, V, and Y are now available to describe any new DNA structure that may appear in the future. However, most of these forms have been created synthetically and have not been observed in naturally occurring biological systems.

There are also triple-stranded DNA forms and quadruplex forms such as the G-quadruplex.

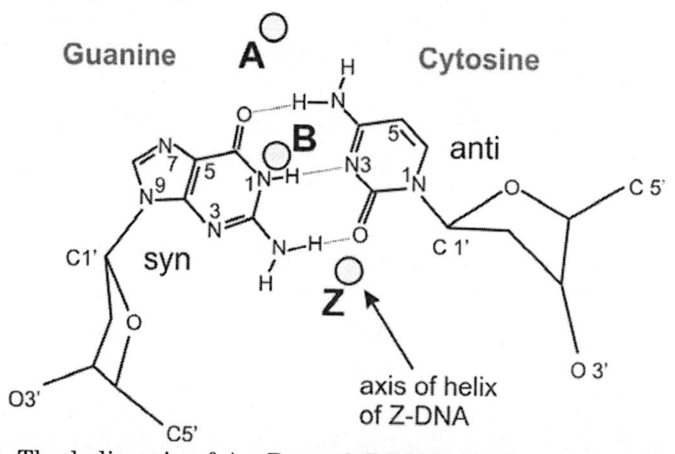


**Figure 7:** The helix axis of A-, B-, and Z-DNA.

**Table 1:** Structural features of the three major forms of DNA

| *Geometry attribute* | *A-DNA* | *B-DNA* | *Z-DNA* |
|---|---|---|---|
| Helix sense | right-handed | right-handed | left-handed |
| Repeating unit | 1 bp | 1 bp | 2 bp |
| Rotation/bp | 32.7° | 35.9° | 60°/2 |
| bp/turn | 11 | 10.5 | 12 |
| Inclination of bp to axis | +19° | "1.2° | "9° |
| Rise/bp along axis | 2.3 E (0.23 nm) | 3.32 E (0.332 nm) | 3.8 E (0.38 nm) |
| Pitch/turn of helix | 28.2 E (2.82 nm) | 33.2 E (3.32 nm) | 45.6 E (4.56 nm) |
| Mean propeller twist | +18° | +16° | 0° |
| Glycosyl angle | anti | anti | C: anti,G: syn |
| Sugar pucker | C3'-endo | C2'-endo | C: C2'-endo,G: C2'-exo |
| Diametre | 23 E (2.3 nm) | 20 E (2.0 nm) | 18 E (1.8 nm) |

### Grooves

Twin helical strands form the DNA backbone. Another double helix may be found by tracing the spaces, or grooves, between the strands. These voids are adjacent to the base pairs and may provide a binding site. As the strands are not directly opposite each other, the grooves are unequally sized.

One groove, the major groove, is 22 E wide and the other, the minor groove, is 12 E wide. The narrowness of the minor groove means that the edges of the bases are more accessible in the major groove. As a result, proteins like transcription factors that can bind to specific sequences in double-stranded DNA usually make contacts to the sides of the bases exposed in the major groove. This situation varies in unusual conformations of DNA within the cell, but the major and minor grooves are always named to reflect the differences in size that would be seen if the DNA is twisted back into the ordinary B form.

### Non-double Helical Forms

Other non-double helical forms of DNA have been theorised, for example the side-by-side (SBS) geometry where the two strands do not wind around each other. Side-by-side models of DNA were proposed early in the history of molecular biology, but these were set aside in favour of the double-helical model due to X-ray crystallography of DNA duplexes and later the nucleosome core particle, as well as the discovery of topoisomerases. The state of current understanding in the field was aptly outlined in an exchange of correspondence in Current Science in 2004..

Single-stranded nucleic acids do not adopt a helical formation, and are described by models such as the random coil or worm-like chain.

### Bending

DNA is a relatively rigid polymer, typically modelled as a worm-like chain. It has three significant degrees of freedom; bending, twisting and compression, each of which cause particular limitations on what is possible with DNA within a cell. Twisting/torsional stiffness is important for the circularisation of DNA and the orientation of DNA bound proteins relative to each other and bending/axial stiffness is important for DNA wrapping and circularisation and protein interactions. Compression/extension is relatively unimportant in the absence of high tension.

### Persistence Length/Axial Stiffness

**Table 2:** Example sequences and their persistence lengths (B DNA)

| ***Sequence*** | ***Persistence Length/base pairs*** |
|---|---|
| Random | 154±10 |
| $(CA)_{repeat}$ | 133±10 |
| $(CAG)_{repeat}$ | 124±10 |
| $(TATA)_{repeat}$ | 137±10 |

DNA in solution does not take a rigid structure but is continually changing conformation due to thermal vibration and collisions with water molecules, which makes classical measures of rigidity impossible. Hence, the bending stiffness of DNA is measured by the persistence length, defined as:

"The length of DNA over which the time-averaged orientation of the polymer becomes uncorrelated by a factor of *e*".

This value may be directly measured using an atomic force microscope to directly image DNA molecules of various lengths. In an aqueous solution, the average persistence length is 46-50 nm or 140-150 base pairs (the diametre of DNA is 2 nm), although can vary significantly. This makes DNA a moderately stiff molecule.

The persistence length of a section of DNA is somewhat dependent on its sequence, and this can cause significant variation. The variation is largely due to base stacking energies and the residues which extend into the minor and major grooves.

### *Models for DNA Bending*

**Table 3:** Stacking stability of base steps (B DNA)

| *Step* | *Stacking $\Delta G$/kcal mol$^{-1}$* |
|---|---|
| T A | -0.19 |
| T G or C A | -0.55 |
| C G | -0.91 |
| A G or C T | -1.06 |
| A A or T T | -1.11 |
| A T | -1.34 |
| G A or T C | -1.43 |
| C C or G G | -1.44 |
| A C or G T | -1.81 |
| G C | -2.17 |

The entropic flexibility of DNA is remarkably consistent with standard polymer physics models such as the *Kratky-Porod* worm-like chain model. Consistent with the worm-like chain model is the observation that bending DNA is also described by Hooke's law at very small (sub-piconewton) forces. However for DNA segments less than the persistence length, the bending force is approximately constant and behaviour deviates from the worm-like chain predictions.

This effect results in unusual ease in circularising small DNA molecules and a higher probability of finding highly bent sections of DNA.

### *Bending Preference*

DNA molecules often have a preferred direction to bend, ie. anisotropic bending. This is, again, due to the properties of the bases which make up the DNA sequence - a random sequence will have no preferred bend direction, i.e. isotropic bending.

Preferred DNA bend direction is determined by the stability of stacking each base on top of the next. If unstable base stacking steps are always found on one side of the DNA helix then the DNA will preferentially bend away from that direction. As bend angle increases then steric hindrances and ability to roll the residues relative to each other also play a role, especially in the minor groove. A and T residues will be preferentially be found in the minor grooves on the inside of bends. This effect is particularly seen in DNA-protein binding where tight DNA bending is induced, such as in nucleosome particles.

DNA molecules with exceptional bending preference can become intrinsically bent. This was first observed in trypanosomatid kinetoplast DNA. Typical sequences which cause this contain stretches of 4-6 T and A residues separated by G and C rich sections which keep the A and T residues in phase with the minor groove on one side of the molecule.

The intrinsically bent structure is induced by the 'propeller twist' of base pairs relative to each other allowing unusual bifurcated Hydrogen-bonds between base steps. At higher temperatures this structure, and so the intrinsic bend, is lost.

All DNA which bends anisotropically has, on average, a longer persistence length and greater axial stiffness. This increased rigidity is required to prevent random bending which would make the molecule act isotropically.

### Circularisation

DNA circularisation depends on both the axial (bending) stiffness and torsional (rotational) stiffness of the molecule. For a DNA molecule to successfully circularise it must be long enough to easily bend into the full circle and must have the correct number of bases so the ends are in the correct rotation to allow bonding to occur. The optimum length for circularisation of DNA is around 400 base pairs (136 nm), with an integral number of turns of the DNA helix, i.e. multiples of 10.4 base pairs. Having a non integral number of turns presents a significant energy barrier for circularisation, for example a 10.4 x 30 = 312 base pair molecule will circularise hundreds of times faster than 10.4 x 30.5 H– 317 base pair molecule.

### Stretching

Longer stretches of DNA are entropically elastic under tension. When DNA is in solution, it undergoes continuous structural variations due to the energy available in the thermal bath of the solvent. This is due to the thermal vibration of the molecule combined with continual collisions with water molecules. For entropic reasons, more compact relaxed states are thermally accessible than stretched out states, and so DNA molecules are almost universally found in a tangled relaxed layouts. For this reason, a single molecule of DNA will stretch under a force, straightening it out. Using optical tweezers, the entropic stretching behaviour of DNA has been studied and analysed from a polymer physics perspective, and it has been found that DNA behaves largely like the *Kratky-Porod* worm-like chain model under physiologically accessible energy scales.

Under sufficient tension and positive torque, DNA is thought to undergo a phase transition with the bases splaying outwards and the phosphates moving to the middle. This proposed structure for overstretched DNA has been called "P-form DNA", in honour of Linus Pauling who originally presented it as a possible structure of DNA.

The mechanical properties DNA under compression have not been characterised due to experimental difficulties in preventing the polymer from bending under the compressive force.

### Supercoiling and Topology

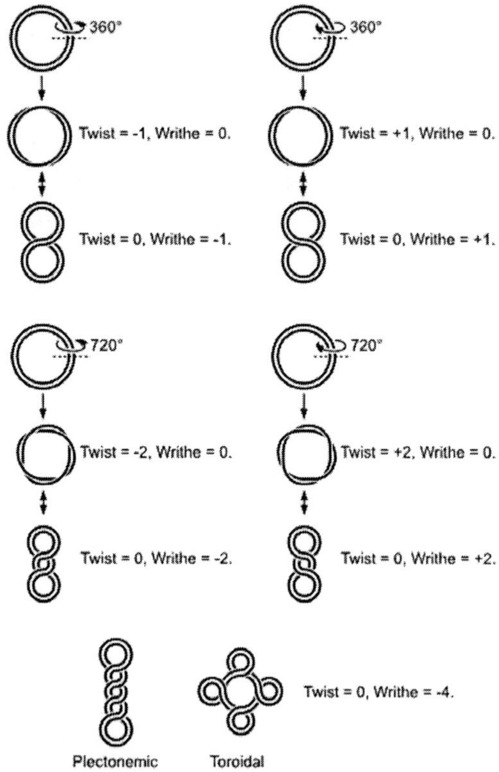


**Figure 8:** Supercoiled structure of circular DNA molecules with low writhe. Helical nature of the DNA duplex is omitted for clarity.

The B form of the DNA helix twists 360° per 10.4-10.5 bp in the absence of torsional strain. But many molecular biological processes can induce torsional strain. A DNA segment with excess or insufficient helical twisting is referred to, respectively, as positively or negatively "supercoiled". DNA *in vivo* is typically negatively supercoiled, which facilitates the unwinding (melting) of the double-helix required for RNA transcription.

Within the cell most DNA is topologically restricted. DNA is typically found in closed loops (such as plasmids in prokaryotes) which are topologically closed, or as very long molecules whose diffusion coefficients produce effectively topologically closed domains. Linear sections of DNA are also commonly bound to proteins or physical structures (such as membranes) to form closed topological loops.

Francis Crick was one of the first to propose the importance of linking numbers when considering DNA supercoils. In a paper published in 1976, Crick outlined the problem as follows:

In considering supercoils formed by closed double-stranded molecules of DNA certain mathematical concepts, such as the linking number and the twist, are needed. The meaning of these for a closed ribbon is explained and also that of the writhing number of a closed curve. Some simple examples are given, some of which may be relevant to the structure of chromatin.

Analysis of DNA topology uses three values:

$L$ = linking number - the number of times one DNA strand wraps around the other. It is an integer for a closed loop and constant for a closed topological domain.

$T$ = twist - total number of turns in the double stranded DNA helix. This will normally tend to approach the number of turns that a topologically open double stranded DNA helix makes free in solution: number of bases/10.5, assuming there are no intercalating agents (e.g., chloroquine) or other elements modifying the stiffness of the DNA.

$W$ = writhe - number of turns of the double stranded DNA helix around the superhelical axis

$$L = T + W \text{ and } \Delta L = \Delta T + \Delta W$$

Any change of T in a closed topological domain must be balanced by a change in W, and vice versa. This results in higher order structure of DNA. A circular DNA molecule with a writhe of 0 will be circular.

If the twist of this molecule is subsequently increased or decreased by supercoiling then the writhe will be appropriately altered, making the molecule undergo plectonemic or toroidal superhelical coiling.

When the ends of a piece of double stranded helical DNA are joined so that it forms a circle the strands are topologically knotted. This means the single strands cannot be separated any process that does not involve breaking a strand (such as heating). The task of unknotting topologically linked strands of DNA falls to enzymes known as topoisomerases. These enzymes are dedicated to unknotting circular DNA by cleaving one or both strands so that another double or single stranded segment can pass through. This unknotting is required for the replication of circular DNA and various types of recombination in linear DNA which have similar topological constraints.

### *The Linking Number Paradox*

For many years, the origin of residual supercoiling in eukaryotic genomes remained unclear. This topological puzzle was referred to by some as the "linking number paradox". However, when experimentally determined structures of the nucleosome displayed an over-twisted left-handed wrap of DNA around the histone octamer, this "paradox" was considered to be solved by the scientific community.

## DNA Nanotechnology

DNA nanotechnology is a branch of nanotechnology which uses the molecular recognition properties of DNA and other nucleic acids to create designed, artificial structures out of DNA for technological purposes. In this field, DNA is used as a structural material rather than as a carrier of genetic information, making it an example of bionanotechnology. DNA nanotechnology has applications in molecular self-assembly and in DNA computing.

Although DNA is usually considered in the context of molecular biology as the carrier of genetic information in living cells, DNA nanotechnology considers DNA solely as a chemical and as a material, and is usually pursued outside of any biological context. DNA nanotechnology makes use of the fact that, due to the specificity of Watson-Crick base pairing, only portions of the strands which are complementary to each other will bind to each other to form duplex DNA. DNA nanotechnology attempts to rationally design sets of DNA strands so that desired portions of each strand will assemble in the correct positions for some desired target structure, a process called nucleic acid design.

Although the field is usually called DNA nanotechnology, its principles apply equally well to other nucleic acids such as RNA and PNA, and structures incorporating these have been made. For this reason the field is occasionally referred to as nucleic acid nanotechnology.

## *Fundamental Concepts*

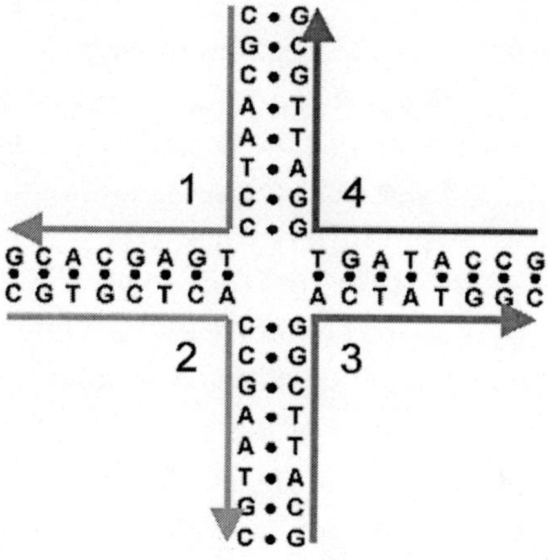


**Figure 9:** These four strands associate into a DNA four-arm junction because this structure maximises the number of correct base pairs, with A's matched to T's and C's matched to G's.

**Figure 10:** A double-crossover (DX) molecule. This molecule consists of five DNA single strands which form two double-helical domains, on the left and the right in this image. There are two crossover points where the strands cross from one domain into the other.

DNA nanotechnology creates complex structures out of nucleic acids by making use of the specificity of base pairing in nucleic acid molecules. The structure of a nucleic acid molecule consists of a sequence of nucleotides, distinguished by which nucleobase they contain. In DNA, the four bases used are adenine(A), ytosine (C), guanine (G), and thymine (T). Nucleic acids have the property that two molecules will bind to each other to form a double helix only if the two sequences are complementary, meaning that they form matching sequences of base pairs, with A's only binding to T's, and C's only to G's.

Because the formation of correctly matched base pairs is energetically favourable, nucleic acid strands are expected in most cases to bind to each other in the conformation that maximises the number of correctly paired bases. This property, that the sequence determines the pattern of binding and the overall structure, is used by the field of DNA nanotechnology in that sequences are artificially designed so that a desired structure is favoured to form.

Nearly all structures in DNA nanotechnology make use of branched DNA structures containing junctions, as opposed to most biological DNA which exists in a linear double helix form. One of the simplest branched structures, and the first made, is a four-arm junction which can be made using four individual DNA strands which are complementary to each other in the correct pattern. Unlike in natural Holliday junctions, in the artificial immobile four-arm junction shown below, the base sequence of each arm is different, meaning that the junction point is fixed in a certain position.

Junctions can be used in more complex molecules. One of the more widely used of these is the "double-crossover" or DX motif. A DX molecule can be thought of as two DNA duplexes positioned parallel to each other, with two crossover points where strands cross from one duplex into the other. Each junction point is itself topologically a four-arm junction. This molecule has the advantage that the junction points are now constrained to a single orientation as opposed to being flexible as in the four-arm junction. This makes the DX motif suitable as a structural building block for larger DNA complexes.

### Design

DNA nanostructures must be rationally designed so that the individual nucleic acid strands will assemble into the desired structures. The design process of such nanostructures usually begins with the specification of a desired target structure and/or functionality. Then, the overall secondary structure of the target molecule is designed, meaning the arrangement of nucleic acid strands within the structure, and which portions of those strands should be bound to each other. The last step is the primary structure design, the specification of the actual base sequences of each nucleic acid strand.

### *Structural Design*

The first step in designing a nucleic acid nanostructure is to decide how a given structure should be represented by a specific arrangement of nucleic acid strands. This design step thus determines the secondary structure, or the series of base pairs which hold the individual strands together in the desired shape. There are several approaches which have been demonstrated:

- Tile-based structures. This approach breaks the target structure into smaller units with strong binding between the strands contained in each unit, and relatively weaker interactions between the units. It is often used to make periodic lattices, but can also be used to implement algorithmic self-assembly, making them one platform for DNA computing.
- Folding structures. An alternative to the tile-based approach, folding approaches make the nanostructure out of a single long strand. This long strand can either have a designed sequence which folds due to its interactions with itself, or it can be folded into the desired shape by using shorter, "staple" strands. This latter method is called DNA origami, which allows the creation of two- and three-dimensional shapes at the nanoscale using DNA.
- Kinetic assembly. Recently, there has been interest in controlling the kinetics of DNA self-assembly, so that transient dynamics can also be programmed into the assembly. Such a method also has the advantage of proceeding isothermally and thus not requiring a thermal annealing step required by solely thermodynamic approaches.

### *Sequence Design*

After any of the above approaches are used to design the secondary structure of a target molecule, an actual sequence of nucleotides must be devised which will form into the desired structure. Nucleic acid design is the process of assigning a specific nucleic acid base sequence to each strand so that they will associate into a desired conformation. Nucleic acid design is central to the field of DNA nanotechnology. Most methods seek to designing sequences so that the target structure is a thermodynamic minimum, and mis-assembled structures have higher energies and are thus disfavoured. This is done either through heuristic methods such as sequence symmetry minimisation and coding theory based approaches, or by explicitly using a full nearest-neighbour thermodynamic model. Geometric models are also used to examine tertiary structure of the nanostructures and ensure that the complexes are not overly strained.

Nucleic acid design has similar goals to protein design: in both, the sequence of monomers is designed to favour the desired folded or associated structure and to disfavour alternate structures. Nucleic acid design has the advantage of being a much computationally simpler problem, since the simplicity of Watson-Crick base pairing rules leads to simple heuristic methods which yield experimentally robust designs. However, nucleic acid structures are less versatile than proteins in their functionality.

### Structural DNA Nanotechnology

Structural DNA nanotechnology, sometimes abbreviated as SDN, focuses on synthesizing and characterising nucleic acid complexes and materials with various nanoscale structures. Structural DNA nanotechnology is largely based on the fact that the three-dimensional structure of DNA—the nucleic acid double helix— has a robust, defined geometry which makes it possible to predict and design the structures of more complex DNA molecules. Many such structures have been created, including two- and three-dimensional structures; and periodic, aperiodic, and discrete structures.

### Periodic Lattices

Smaller nucleic acid assemblies can be equipped with sticky ends in order to combine them into a two-dimensional periodic lattice. The earliest example of this was the array of DX, or double-crossover, molecules. Each DX molecule can be designed with four sticky ends, one at each end of the two double-helical domains, and these sticky ends can be designed with sequences that cause the DX units to combine into a specific tessellated pattern. They thus form extended flat sheets which are essentially rigid two-dimensional crystals of DNA.

Two-dimensional arrays have been made out of other motifs as well, including the Holliday junction rhombus array as well as various DX-based arrays making use of a double-cohesion scheme.

Creating three-dimensional lattices out of DNA was the earliest goal of DNA nanotechnology, but proved to be one of the most difficult to realise. Success in constructing three-dimensional DNA lattices was finally reported in 2009 using a motif based on the concept of tensegrity, a balance between tension and compression forces.

### *Nanotubes*

In addition to flat sheets, DX arrays have been made to form hollow nanotubes of 4–20 nm diametre. These DNA nanotubes are somewhat similar in size and shape to carbon nanotubes, but the carbon nanotubes are stronger and better conductors, whereas the DNA nanotubes are more easily modified and connected to other structures. There have been multiple schemes for constructing DNA nanotubes, one of which uses the inherent curvature of DX tiles to form a DX lattice to curl around itself and close into a tube. An alternative design uses single-stranded “tiles” for which the rigidity of the tube is an emergent property. This method also has the benefit of being able to determine the circumference of the nanotube in a simple, modular fashion.

### *Polyhedra*

A number of three-dimensional DNA molecules have been made which have the connectivity of a polyhedron such as an octahedron or cube. In other words, the DNA duplexes trace the edges of a polyhedron with a DNA junction at each vertex. The earliest demonstrations of DNA polyhedra involved multiple ligations and solid-phase synthesis steps to create catenated polyhedra. More recent work has yielded polyhedra whose synthesis is much easier. These include a DNA octahedron made from a long single strand designed to fold into the correct conformation, as well as a tetrahedron which can be produced from four DNA strands in a single step, pictured at the top of this article.

### *Arbitrary Shapes*

Nanostructures of arbitrary shapes are usually made using the DNA origami method. DNA origami makes use of a long natural virus strand as a “scaffold” strand, and computationally designs shorter “staple” strands which bind to portions of the scaffold strand and force it to fold into the desired shape. This method has the advantage of being easy to design, as the base sequence is predetermined by the scaffold strand sequence, and it also does not require high strand purity and accurate stoichiometry, as most other DNA nanotechnology methods do. DNA origami was first demonstrated for two-dimensional shapes; demonstrated designs included the smiley face and a coarse map of North America. This was later extended to solid three-dimensional shapes.

In addition, structures have been constructed with two-dimensional faces which fold into an overall three-dimensional shape, akin to a cardboard box. These can be programmed to open and release their cargo in response to a stimulus, making them potentially useful as programmable molecular cages.

### *Functional Nucleic Acid Nanostructures*

DNA nanotechnology focuses on creating molecules with designed functionalities as well as structures. These include both dynamic functionality within the nucleic acid structure itself, for example with computation and mechanical motion, as well as by including other components such as small molecules or nanoparticles which have their own functionalities. Many classes of functional systems have been demonstrated.

### *Nanoarchitecture*

The idea of using DNA arrays to template the assembly of other functional molecules was first suggested by Nadrian Seeman in 1987, but only recently has progress been made in reducing these kinds of schemes to practice.

In 2006, researchers covalently attached gold nanoparticles to a DX-based tile and showed that self-assembly of the DNA structures also assembled the nanoparticles hosted on them. Also that year, Dwyer and LaBean demonstrated the letters "D" "N" and "A" created on a 4x4 DX array using streptavidin, and a hierarchical assembly based on this approach was also demonstrated that scales to larger arrays (8X8 and 8.96 MD). A non-covalent hosting scheme was shown in 2007, using Dervan polyamides on a DX array to arrange streptavidin proteins on specific kinds of tiles on the DNA array.

There has also been interest in using DNA nanotechnology to assemble molecular electronics devices. To this end, DNA has been used to assemble single walled carbon nanotubes into field-effect transistors.

## *Algorithmic Self-assembly*

**Figure 11:** The Sierpinski gasket fractal.

**Figure 12:** DNA arrays that display a representation of the Sierpinski gasket on their surfaces.

DNA nanotechnology has been applied to the related field of DNA computing. The DX tiles can have their sticky end sequences chosen so that they act as Wang tiles, allowing them to perform computation. A DX array has been demonstrated whose assembly encodes an XOR operation; this allows the DNA array to implement a cellular automaton which generates a fractal called the Sierpinski gasket. Another system has the function of a binary counter, displaying a representation of increasing binary numbers as it grows. These results show that computation can be incorporated into the assembly of DNA arrays, increasing its scope beyond simple periodic arrays. DNA computing overlaps with, but is distinct from, DNA nanotechnology. The latter uses the specificity of Watson-Crick basepairing to make novel structures out of DNA. These structures can be used for DNA computing, but they do not have to be. Additionally, DNA computing can be realised without using the types of molecules made possible by DNA nanotechnology.

### Nanomechanical Devices

DNA complexes have been made which change their conformation upon some stimulus. These are intended to have applications in nanorobotics. DNA machines have also been made which show a twisting motion. The first such device made use of the transition between the B-DNA and Z-DNA forms to respond to a change in buffer conditions. This reliance on buffer conditions, however, caused all devices to change state at the same time. A subsequent system, called "molecular tweezers," changes from an open to a closed state based upon the presence of control strands, allowing multiple devices to be individually operated in solution. This was followed up by another system which relies on the presence of control strands to switch from a paranemic-crossover (PX) conformation to a double-junction (JX2) conformation.

Nucleic acid nanomachines have been made which exhibit directional motion along a linear track, called DNA walkers. A large number of schemes have been demonstrated. One strategy is to control the motion of the walker along the track using control strands which need to be manually added in sequence. Another approach is to make use of restriction enzymes or deoxyribozymes to cleave the strands and cause the walker to move forward, which has the advantage of running autonomously. A later system extended the concept of DNA walkers to walk upon a two-dimensional surface rather than a linear track, and demonstrated the ability to selectively pick up and move molecular cargo. Additionally, a linear walker has been demonstrated which performs DNA-templated synthesis as the walker advances along the track, allowing autonomous multistep chemical synthesis directed by the walker.

### Applications

DNA nanotechnology provides one of the only ways to form designed, complex structures with precise control over nanoscale features. The field is beginning to see application to solvebasic science problems in structural biology and biophysics. One such application, the earliest envisaged for the field, is in crystallography, where molecules which are hard to crystallise by themselves could be arranged and oriented within a three-dimensional nucleic acid lattice, thus allowing determination of their structure. DNA origami rods have also been used to replace liquid crystals in residual dipolar coupling experiments in protein NMR spectroscopy; using DNA origami is advantageous because, unlike liquid crystals, they are tolerant of the detergents needed to suspend membrane proteins in solution. It has been demonstrated that DNA walkers can be used as nanoscale assembly lines to move nanoparticles and direct chemical synthesis. Furthermore, DNA origami structures have aided in the biophysical studies of enzyme function and protein folding.

DNA nanotechnology is also moving towards potential real-world applications. It has been suggested that the ability of nucleic acid arrays to arrange other molecules has potential applications in molecular scale electronics, with the assembly of a nucleic acid lattice templating the assembly of molecular electronic elements such as molecular wires. DNA nanotechnology has also been called a form of programmable matter because of the coupling of computation to its material properties.

There are also potential applications for DNA nanotechnology in nanomedicine, making use of its ability to perform computation in a biocompatible format to make "smart drugs" for targeted drug delivery. One such system being investigated uses a hollow DNA box containing proteins which induce apoptosis, or cell death, which will only open when in proximity to a cancer cell.

### *Materials and Methods*

The sequences of the individual DNA strands which make up the target structure are designed computationally, using molecular modeling and thermodynamic modeling software. Once the sequences have been designed, the nucleic acid molecules themselves are synthesized through standard oligonucleotide synthesis methods. This process is usually automated by using a machine called an oligonucleotide synthesizer, and nucleic acids of custom sequence are commercially available from many vendors. For methods which require pure strands of known concentration, the nucleic acid strands can be purified using denaturing gel electrophoresis, and concentrations are determined by one of several nucleic acid quantitation methods using ultraviolet absorbance spectroscopy.

The fully formed target structures are usually characterised by native gel electrophoresis, which provides information about the size and shape of DNA molecules. An electrophoretic mobility shift assay can indicate whether a structure incorporates all the individual desired strands. Fluorescent labelling and Fϕrster resonance energy transfer (FRET) are also used to characterise the structure of the molecules.

Nucleic acid structures can be directly imaged by atomic force microscopy, which is well-suited to extended two-dimensional structures, but is less useful for discrete three-dimensional structures due to the interaction of the microscope tip with the fragile nucleic acid structure. For these latter structures transmission electron microscopy and cryo-electron microscopy are important methods. Extended three-dimensional lattices are analysed by X-ray crystallography.

### History

**Figure 13:** The M. C. Escher woodcut *Depth*(pictured) inspired Nadrian Seeman to consider using three-dimensional lattices of DNA to orient hard-to-crystallise molecules. This led to the beginning of the field of DNA nanotechnology.

The concept of DNA nanotechnology was invented by Nadrian Seeman in the early 1980s. Seeman was originally concerned with using a three-dimensional DNA lattice to orient target molecules, which would simplify their crystallographic study by eliminating the difficult process of obtaining pure crystals. This idea had reportedly come to him in fall 1980, after realising the similarity between the M. C. Escher woodcut *Depth* and an array of DNA six-arm junctions.

To this end, Seeman's laboratory in 1991 published the synthesis of a cube made of DNA, the first three-dimensional nanoscale object, for which he received the 1995 Feynman Prize in Nanotechnology, which was followed by a DNA truncated octahedron. However, it soon became clear that these molecules, polygonal shapes with flexible junctions as their vertices, were not rigid enough to form extended three-dimensional lattices. Seeman developed the more rigid double-crossover (DX) motif, and in collaboration with Erik Winfree, in 1998 published the creation of two-dimensional lattices of DX tiles. The synthesis of a three-dimensional lattice was finally published by Seeman in 2009, nearly thirty years after he had set out to do so.

These tile-based structures had the advantage that they provided the capability to implement DNA computing, which was demonstrated by Winfree and Paul Rothemund in their 2004 paper on the algorithmic self-assembly of a Sierpinski gasket structure, and for which they shared the 2006 Feynman Prize in Nanotechnology. Winfree's key insight was that the DX tiles could be used as Wang tiles, meaning that their assembly was capable of performing computation.

The first DNA nanomachine—a motif which changes its structure in response to an input—was demonstrated in 1999 by Seeman. An improved system was demonstrated by Bernard Yurke the following year, which was also the first nucleic acid device to make use of toehold-mediated strand displacement. The next advance was to translate this into mechanical motion, and in 2004 and 2005, a number of DNA walker systems were demonstrated by the groups of Seeman, Niles Pierce, Andrew Turberfield, and Chengde Mao.

In 2006, Rothemund first demonstrated the new DNA origami technique for easily and robustly creating folded DNA molecules of any shape. Rothemund had conceived of this method as being conceptually intermediate between Seeman's DX lattices, which used many short strands, and William Shih's DNA octahedron, which consisted mostly of one very long strand; Rothemund's DNA origami contains a long strand whose folding is assisted by a number of short strands. This method allowed the creation of much larger structures than were previously possible, which are less technically demanding to design and synthesize. DNA origami was the cover story of *Nature* on March 15, 2006. Rothemund's research demonstrating two-dimensional DNA origami structures was followed by the demonstration of solid three-dimensional DNA origami by Shih in 2009, while the labs of Jørgen Kjems and Hao Yan demonstrated hollow three-dimensional structures made out of two-dimensional faces.

DNA nanotechnology as a field was initially met with some skepticism due to the unusual non-biological use of nucleic acids as a material for building structures and doing computation, as well as the preponderance of proof of principle experiments which extended the capabilities of the field but were far from actual applications. Seeman's seminal 1991 paper on the synthesis of the DNA cube was rejected by the journal *Science* after one reviewer praised its originality while another criticized it for its lack of biological relevance. By the early 2010s, however, the field was considered to have increased its capabilities to the point that applications for basic science research were beginning to be realised, and practical applications in medicine and other fields were beginning to be considered feasible. In addition, the field had grown from very few active laboratories in 2001, to at least 60 in 2010, which increased the talent pool and thus the number of advancements in the field during that decade.

## Molecular Models of DNA

Molecular models of DNA structures are representations of the molecular geometry and topology of Deoxyribonucleic acid (DNA) molecules using one of several means, with the aim of simplifying and presenting the essential, physical and chemical, properties of DNA molecular structures either *in vivo* or *in vitro*. These representations include closely packed spheres (CPK models) made of plastic, metal wires for 'skeletal models', graphic computations and animations by computers, artistic rendering. Computer molecular models also allow animations and molecular dynamics simulations that are very important for understanding how DNA functions *in vivo*.

The more advanced, computer-based molecular models of DNA involve molecular dynamics simulations as well as quantum mechanical computations of vibro-rotations, delocalised molecular orbitals (MOs), electric dipole moments, hydrogen-bonding, and so on. DNA molecular dynamics modeling involves simulations of DNA molecular geometry and topology changes with time as a result of both intra- and inter- molecular interactions of DNA. Whereas molecular models of Deoxyribonucleic acid (DNA) molecules such as closely packed spheres (CPK models) made of plastic or metal wires for 'skeletal models' are useful representations of static DNA structures, their usefulness is very limited for representing complex DNA dynamics. Computer molecular modeling allows both animations and molecular dynamics simulations that are very important for understanding how DNA functions *in vivo*.

### *History*

From the very early stages of structural studies of DNA by X-ray diffraction and biochemical means, molecular models such as the Watson-Crick double-helix model were successfully employed to solve the 'puzzle' of DNA structure, and also find how the latter relates to its key functions in living cells. The first high quality X-ray diffraction patterns of A-DNA were reported by Rosalind Franklin and Raymond Gosling in 1953. The first calculations of the Fourier transform of an atomic helix were reported one year earlier by Cochran, Crick and Vand, and were followed in 1953 by the computation of the Fourier transform of a coiled-coil by Crick.

Structural information is generated from X-ray diffraction studies of oriented DNA fibers with the help of molecular models of DNA that are combined with crystallographic and mathematical analysis of the X-ray patterns.

The first reports of a double-helix molecular model of B-DNA structure were made by Watson and Crick in 1953. Last-but-not-least, Maurice F. Wilkins, A. Stokes and H.R. Wilson, reported the first X-ray patterns of *in vivo* B-DNA in partially oriented salmon sperm heads. The development of the first correct double-helix molecular model of DNA by Crick and Watson may not have been possible without the biochemical evidence for the nucleotide base-pairing ([A—T]; [C—G]), or Chargaff's rules. Although such initial studies of DNA structures with the help of molecular models were essentially static, their consequences for explaining the *in vivo* functions of DNA were significant in the areas of protein biosynthesis and the quasi-universality of the genetic code. Epigenetic transformation studies of DNA *in vivo* were however much slower to develop in spite of their importance for embryology, morphogenesis and cancer research. Such chemical dynamics and biochemical reactions of DNA are much more complex than the molecular dynamics of DNA physical interactions with water, ions and proteins/enzymes in living cells.

### Importance

An old standing dynamic problem is how DNA "self-replication" takes place in living cells that should involve transient uncoiling of supercoiled DNA fibers. Although DNA consists of relatively rigid, very large elongated biopolymer molecules called "fibers" or chains (that are made of repeating nucleotide units of four basic types, attached to deoxyribose and phosphate groups), its molecular structure *in vivo* undergoes dynamic configuration changes that involve dynamically attached water molecules and ions.

Supercoiling, packing with histones in chromosome structures, and other such supramolecular aspects also involve *in vivo* DNA topology which is even more complex than DNA molecular geometry, thus turning molecular modeling of DNA into an especially challenging problem for both molecular biologists and biotechnologists.

Like other large molecules and biopolymers, DNA often exists in multiple stable geometries (that is, it exhibits conformational isomerism) and configurational, quantum states which are close to each other in energy on the potential energy surface of the DNA molecule.

Such varying molecular geometries can also be computed, at least in principle, by employing *ab initio* quantum chemistry methods that can attain high accuracy for small molecules, although claims that acceptable accuracy can be also achieved for polynuclelotides, as well as DNA conformations, were recently made on the basis of VCD spectral data.

Such quantum geometries define an important class of *ab initio* molecular models of DNA whose exploration has barely started especially in connection with results obtained by VCD in solutions. More detailed comparisons with such *ab initio* quantum computations are in principle obtainable through 2D-FT NMR spectroscopy and relaxation studies of polynucleotide solutions or specifically labelled DNA, as for example with deuterium labels.

In an interesting twist of roles, the DNA molecule itself was proposed to be utilised for quantum computing. Both DNA nanostructures as well as DNA 'computing' biochips have been built.

### *Fundamental Concepts*

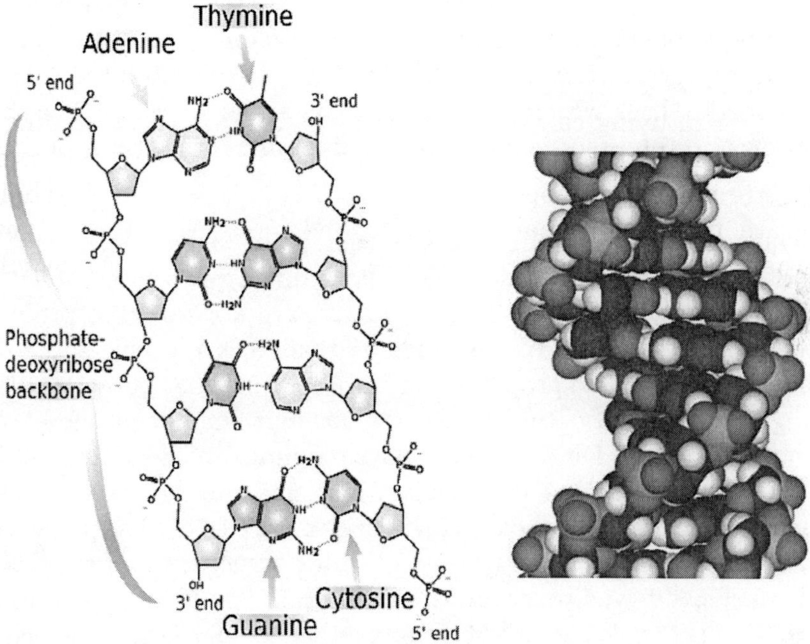


**Figure 14:** At left, the chemical structure of DNA showing the base-pairing. This depiction of a DNA duplex lacks information about the molecule's three-dimensional structure, at right.

The chemical structure of DNA is insufficient to understand the complexity of the 3D structures of DNA. On the other hand, animated molecular models allow one to visually explore the three-dimensional (3D) structure of DNA. The DNA model shown (far right) is a space-filling, or CPK, model of the DNA double-helix. Animated molecular models, such as the wire, or skeletal, type shown at the top of this article, allow one to visually explore the three-dimensional (3D) structure of DNA. Another type of DNA model is the space-filling, or CPK, model.

The hydrogen bonding dynamics and proton exchange is very different by many orders of magnitude between the two systems of fully hydrated DNA and water molecules in ice. Thus, the DNA dynamics is complex, involving nanosecond and several tens of picosecond time scales, whereas that of liquid ice is on the picosecond time scale, and that of proton exchange in ice is on the millisecond time scale. The proton exchange rates in DNA and attached proteins may vary from picosecond to nanosecond, minutes or years, depending on the exact locations of the exchanged protons in the large biopolymers.

A simple harmonic oscillator 'vibration' is only an oversimplified dynamic representation of the longitudinal vibrations of the DNA intertwined helices which were found to be an harmonic rather than harmonic as often assumed in quantum dynamic simulations of DNA.

### DNA Structure

The structure of DNA shows a variety of forms, both double-stranded and single-stranded. The mechanical properties of DNA, which are directly related to its structure, are a significant problem for cells. Every process which binds or reads DNA is able to use or modify the mechanical properties of DNA for purposes of recognition, packaging and modification. The extreme length (a chromosome may contain a 10 cm long DNA strand), relative rigidity and helical structure of DNA has led to the evolution of histones and of enzymes such as topoisomerases and helicases to manage a cell's DNA. The properties of DNA are closely related to its molecular structure and sequence, particularly the weakness of the hydrogen bonds and electronic interactions that hold strands of DNA together compared to the strength of the bonds within each strand.

Experimental techniques which can directly measure the mechanical properties of DNA are relatively new, and high-resolution visualisation in solution is often difficult. Nevertheless, scientists have uncovered large amount of data on the mechanical properties of this polymer, and the implications of DNA's mechanical properties on cellular processes is a topic of active current research.

The DNA found in many cells can be macroscopic in length - a few centimetres long for each human chromosome. Consequently, cells must compact or "package" DNA to carry it within them. In eukaryotes this is carried by spool-like proteins known as histones, around which DNA winds. It is the further compaction of this DNA-protein complex which produces the well known mitotic eukaryotic chromosomes.

Alternate non-double-helical models have been proposed for the *in vivo* structure of DNA, such as in plasmids and chromatin. Such "side-by-side" models of DNA were proposed early in the history of molecular biology, but these were set aside in favour of the double-helical model due to X-ray crystallography of DNA duplexes and later the nucleosome core particle, as well as the discovery of topoisomerases, and these non-double-helical models are not currently accepted by the mainstream scientific community.

## DNA Structure Determination using Molecular Modeling and DNA X-ray Patterns

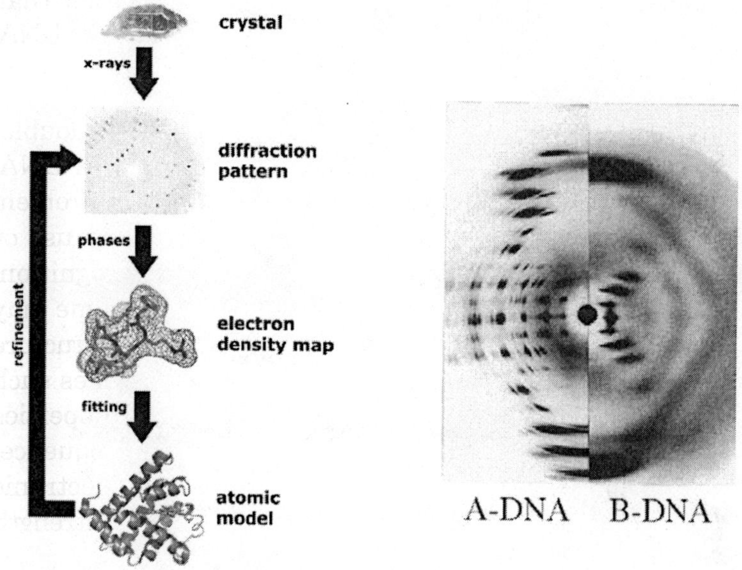


**Figure 15:** Left, the major steps involved in DNA structure determination by X-ray crystallography showing the important role played by molecular models of DNA structure in this iterative process. Right, an image of actual A- and B- DNA X-ray patterns obtained from oriented and hydrated DNA fibers.

After DNA has been separated and purified by standard biochemical techniques one has a sample in a jar. Below are the main steps involved in generating structural information from X-ray diffraction studies of oriented DNA fibers that are drawn from the hydrated DNA sample with the help of molecular models of DNA that are combined with crystallographic and mathematical analysis of the X-ray patterns.

## Paracrystalline Lattice Models of B-DNA Structures

A paracrystalline lattice, or paracrystal, is a molecular or atomic lattice with significant amounts (e.g., larger than a few percent) of partial disordering of molecular arrangements. Limiting cases of the paracrystal model are nanostructures, such as glasses, liquids, etc., that may possess only local ordering and no global order.

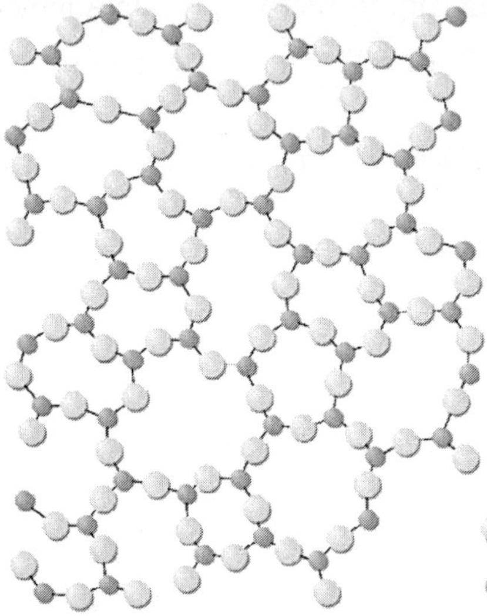


**Figure 16:** Silica glass is another example of a material which is organised into a paracrystalline lattice.

Liquid crystals also have paracrystalline rather than crystalline structures.

Highly hydrated B-DNA occurs naturally in living cells in such a paracrystalline state, which is a dynamic one in spite of the relatively rigid DNA double-helix stabilised by parallel hydrogen bonds between the nucleotide base-pairs in the two complementary, helical DNA chains. For simplicity most DNA molecular models omit both water and ions dynamically bound to B-DNA, and are thus less useful for understanding the dynamic behaviours of B-DNA *in vivo*. The physical and mathematical analysis of X-ray and spectroscopic data for paracrystalline B-DNA is therefore much more complicated than that of crystalline, A-DNA X-ray diffraction patterns. The paracrystal model is also important for DNA technological applications such as DNA nanotechnology. Novel techniques that combine X-ray diffraction of DNA with X-ray microscopy in hydrated living cells are now also being developed.

### *Genomic and Biotechnology Applications of DNA Molecular Modeling*

There are various uses of DNA molecular modeling in Genomics and Biotechnology research applications, from DNA repair to PCR and DNA nanostructures. Two-dimensional DNA junction arrays have been visualised by Atomic force microscopy. DNA molecular modeling has various uses in genomics and biotechnology, with research applications ranging from DNA repair to PCR and DNA nanostructures.

These include computer molecular models of molecules as varied as RNA polymerase, an E. coli, bacterial DNA primase template suggesting very complex dynamics at the interfaces between the enzymes and the DNA template, and molecular models of the mutagenic, chemical interaction of potent carcinogen molecules with DNA. These are all represented in the gallery below.

Technological application include a DNA biochip and DNA nanostructures designed for DNA computing and other dynamic applications of DNA nanotechnology. The image at right is of self-assembled DNA nanostructures.

The DNA "tile" structure in this image consists of four branched junctions oriented at 90° angles. Each tile consists of nine DNA oligonucleotides as shown; such tiles serve as the primary "building block" for the assembly of the DNA nanogrids shown in the AFM micrograph. Quadruplex DNA may be involved in certain cancers. Images of quadruplex DNA are in the gallery below.

## Molecular Structure of Nucleic Acids: A Structure for Deoxyribose Nucleic Acid

The "Molecular structure of Nucleic Acids: A Structure for Deoxyribose Nucleic Acid" was an article published by James D. Watson and Francis Crick in the scientific journal *Nature* in its 171st volume on pages 737–738 (dated April 25, 1953). It was the first publication which described the discovery of the double helix structure of DNA.

This discovery had a major impact on biology, particularly in the field of genetics.

This article is often termed a "pearl" of science because it is brief and contains the answer to a fundamental mystery about living organisms. This mystery was the question of how it was possible that genetic instructions were held inside organisms and how they were passed from generation to generation. The article presents a simple and elegant solution, which surprised many biologists at the time who believed that DNA transmission was going to be more difficult to detail and understand.

## The Nature of the Discovery

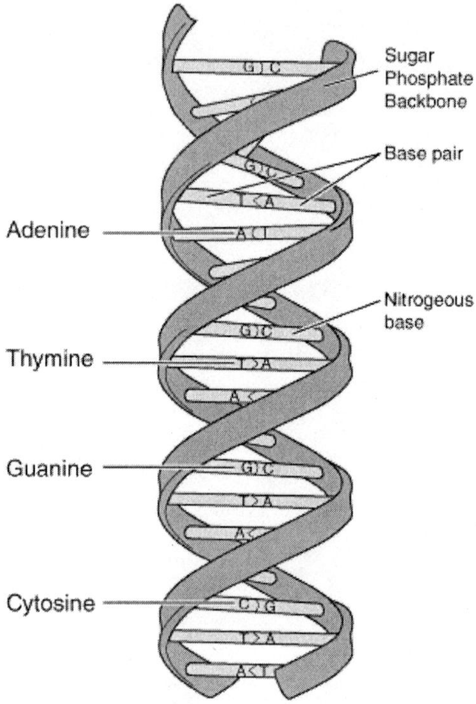


**Figure 17:** Diagramatic representation of the key structural features of the DNA double helix.

Watson and Crick's 1953 article contains the answer to a fundamental mystery about living organisms. The nature of their discovery was distinctive and in some ways unsurprising. What is hidden in the technical jargon of the title is that Watson and Crick's discovery of the chemical structure of DNA finally revealed how genetic instructions are stored inside organisms and passed from generation to generation.

## Origins of Molecular Biology

The application of physics and chemistry to biological problems led to the development of molecular biology. Not all biology that concerns molecules falls into the category that is labelled "molecular biology". Molecular biology is particularly concerned with the flow and consequences of biological information at the level of genes and proteins. The discovery of the DNA double helix made clear that genes are functionally defined parts of DNA molecules and that there must be a way for cells to make use of their DNA genes in order to make proteins.

Linus Pauling was a chemist who was very influential in developing an understanding of the structure of biological molecules. In 1951, Pauling published the structure of the alpha helix, a fundamentally important structural component of proteins. Early in 1953 Pauling published an incorrect triple helix model of DNA. Both Crick, and particularly Watson, thought that they were racing against Pauling to discover the structure of DNA.

Max Delbrück was a physicist who recognised some of the biological implications of quantum physics. Delbruck's thinking about the physical basis of life stimulated Erwin Schrϕdinger to write the highly influential book, *What Is Life?* Schrϕdinger's book was an important influence on Francis Crick, James D. Watson, and Maurice Wilkins who won a Nobel prize in Medicine for the discovery of the DNA double helix. Delbruck's efforts to promote the "Phage Group" (exploring genetics by way of the viruses that infect bacteria) was important in the early development of molecular biology in general and the development of Watson's scientific interests in particular.

### *DNA Structure and Function*

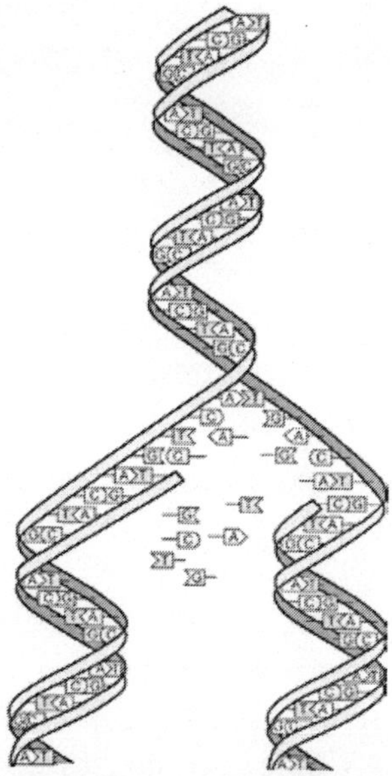

**Figure 18:** DNA replication. The two base-pair complementary chains of the DNA molecule allow for replication of the genetic instructions.

It is not always the case that the structure of a molecule is easy to relate to its function. What makes the structure of DNA so obviously related to its function was described modestly at the end of the article: "It has not escaped our notice that the specific pairing we have postulated immediately suggests a possible copying mechanism for the genetic material".

The "specific pairing" is a key feature of the Watson and Crick model of DNA, the pairing of nucleotide subunits. In DNA, the amount of guanine is equal to cytosine and the amount of adenine is equal to thymine. The A:T and C:G pairs are structurally similar. In particular, the length of each base pair is the same and they fit equally between the two phosphate backbones. The base pairs are held together by hydrogen bonds, a type of chemical attraction that is easy to break and easy to reform. After realising the structural similarity of the A:T and C:G pairs, Watson and Crick soon produced their double helix model of DNA with the hydrogen bonds at the core of the helix providing a way to unzip the two complementary strands for easy replication: the last key requirement for a likely model of the genetic molecule.

Indeed, the base-pairing did suggest a way to copy a DNA molecule. Just pull apart the two phosphate backbones, each with its hydrogen bonded A, T, G, and C components. Each strand could then be used as a template for assembly of a new base-pair complementary strand.

### *Future Considerations*

When Watson and Crick produced their double helix model of DNA, it was known that most of the specialised features of the many different life forms on Earth are made possible by proteins. Structurally, proteins are long chains of amino acid subunits. In some way, the genetic molecule, DNA, had to contain instructions for how to make the thousands of proteins found in cells.

From the DNA double helix model, it was clear that there must be some correspondence between the linear sequences of nucleotides in DNA molecules to the linear sequences of amino acids in proteins. The details of how sequences of DNA instruct cells to make specific proteins was worked out by molecular biologists during the period from 1953 to 1965. Francis Crick played an integral role in both the theory and analysis of the experiments that led to an improved understanding of the genetic code.

### Consequences

Other advances in molecular biology stemming from the discovery of the DNA double helix eventually led to ways to sequence genes. James Watson played an important role in getting government funding for the Human Genome Project. The ability to sequence and manipulate DNA is now central to the biotechnology industry and modern medicine. The austere beauty of the structure and the practical implications of the DNA double helix combined to make *Molecular structure of Nucleic Acids; A Structure for Deoxyribose Nucleic Acid* one of the most prominent biology articles of the twentieth century.

### Controversy

Watson and Crick based their molecular model of the DNA double helix on data that had been collected by researchers in several other laboratories. Watson and Crick were the first to put together all of the scattered fragments of information that were required to produce a successful molecular model of DNA.

Much of the data that were used by Crick and Watson came from unpublished work by Maurice Wilkins, Rosalind Franklin, A.R. Stokes, and H.R. Wilson at King's College London in the University of London. Key data from Wilkins, Stokes, and Wilson, and, separately, by Franklin and Gosling, were published in two separate additional articles in the same issue of *Nature* with the article by Watson and Crick. The article by Watson and Crick did acknowledge that they had been "stimulated" by experimental results from the King's College researchers, and a similar acknowledgment was published by M. H. F. Wilkins, A.R. Stokes, and H. R. Wilson in the following three-page article.

In 1968, Watson published a highly controversial autobiographical account of the discovery of the double-helical, molecular structure of DNA called *The Double Helix*, and which was not accepted—at least publicly—either by Francis Crick or by M.H.F. Wilkins. Furthermore, Erwin Chargaff also printed a rather "unsympathetic review" of James D. Watson's booklet in the March 29, 1968 issue of *Science*. In his `autobiographical' booklet, Watson stated among other things that he and Crick had access to some of Franklin's data from a source that she was not aware of, and also that he had seen—without her permission—the B-DNA X-ray diffraction pattern obtained by Franklin and Gosling in May 1952 at King's in London. In particular, in late 1952, Dr. Franklin had submitted a progress report to the Medical Research Council, which was reviewed by Dr. Max Perutz, then at The Cavendish Laboratory of the University of Cambridge, UK. Watson and Crick also worked in the MRC-supported Cavendish Laboratory in Cambridge whereas Drs. Wilkins and Franklin were in the MRC supported laboratory at King's in London. Such MRC reports were not usually widely circulated, but Crick read a copy of Dr. Franklin's research summary in early 1953.

Max Perutz's justification for passing this information to both Crick and Watson was that the report contained information which Watson has previously heard in November 1951 when Dr. Franklin talked about her unpublished results with Raymond Gosling during a meeting arranged by Dr. M.H.F. Wilkins at King's College, following a request from Crick and Watson; this justification does not hold however for Crick who was not present at this November 1951 meeting, but who also was given access by Max Perutz to Franklin's MRC report data which prompted Crick and Watson to seek permission from Sir Lawrence Bragg—who was at the time the head of the Cavendish Laboratory in Cambridge—to publish in *Nature* their double-helix molecular model of DNA based on Franklin's and also Wilkins's data.

Moreover, in November 1951 Watson had acquired—by his own admission—little training in X-ray crystallography, and therefore had not fully understood (again, according to his own admission, in "The Double Helix") what Dr. Franklin was saying about the structural symmetry of the DNA molecule. Crick, however, knowing the Fourier transforms of Bessel functions that represent the X-ray diffraction patterns of helical structures of atoms, correctly interpreted further one of Dr.Franklin's experimental findings as indicating that DNA was most likely to be a double helix with the two polynucleotide chains running in opposite directions.

Crick was thus in a unique position to make this interpretation because he had previously worked on the X-ray diffraction data for other large molecules that had similar, helical symmetry to that of DNA. Dr. Franklin, on the other hand, rejected at first the molecular model building approach proposed by Crick and Watson because their first DNA model presented by Watson to her and Dr. M.H.F. Wilkins in 1952 in London had an obviously incorrect structure with hydrated charged groups on the inside of the model, rather than on the outside, as explicitly admitted by James D. Watson in his "Double Helix" booklet. It is therefore questioned whether Crick's colleague, Dr.Max Perutz, acted unethically by allowing Crick access to Dr. Franklin's MRC report about the crystallographic unit of the B-DNA and A-DNA structures. Dr. Perutz claimed, however, that he believed he had not because this report was not confidential, and had been designed as part of an effort to promote contact between different MRC research groups.

## Crystallography

Crystallography is the experimental science of the arrangement of atoms in solids. The word "crystallography" derives from the Greek words *crystallon* = cold drop / frozen drop, with its meaning extending to all solids with some degree of transparency, and *grapho* = write.

Before the development of X-ray diffraction crystallography, the study of crystals was based on their geometry. This involves measuring the angles of crystal faces relative to theoretical reference axes (crystallographic axes), and establishing the symmetry of the crystal in question. The former is carried out using a goniometre. The position in 3D space of each crystal face is plotted on a stereographic net, e.g. Wulff net or Lambert net. In fact, the pole to each face is plotted on the net. Each point is labelled with its Miller index. The final plot allows the symmetry of the crystal to be established.

Crystallographic methods now depend on the analysis of the diffraction patterns of a sample targeted by a beam of some type. Although X-rays are most commonly used, the beam is not always electromagnetic radiation. For some purposes electrons or neutrons are used. This is facilitated by the wave properties of the particles. Crystallographers often explicitly state the type of illumination used when referring to a method, as with the terms *X-ray diffraction, neutron diffraction* and *electron diffraction.*

These three types of radiation interact with the specimen in different ways. X-rays interact with the spatial distribution of the valence electrons, while electrons are charged particles and therefore feel the total charge distribution of both the atomic nuclei and the surrounding electrons. Neutrons are scattered by the atomic nuclei through the strong nuclear forces, but in addition, the magnetic moment of neutrons is non-zero. They are therefore also scattered by magnetic fields. When neutrons are scattered from hydrogen-containing materials, they produce diffraction patterns with high noise levels. However, the material can sometimes be treated to substitute hydrogen for deuterium. Because of these different forms of interaction, the three types of radiation are suitable for different crystallographic studies.

### Theory

Generally, an image of a small object is made using a lens to focus the illuminating radiation, as is done with the rays of the visible spectrum in light microscopy. However, the wavelength of visible light (about 4000 to 7000 angstroms) is three orders of magnitude longer than the length of typical atomic bonds and atoms themselves (about 1 to 2 angstroms). Therefore, obtaining information about the spatial arrangement of atoms requires the use of radiation with shorter wavelengths, such as X-rays or neutrons beam. Employing shorter wavelengths implied abandoning microscopy and true imaging, however, because there exists no material from which a lens capable of focusing this type of radiation can be created. (That said, scientists have had some success focusing X-rays with microscopic Fresnel zone plates made from gold, and by critical-angle reflection inside long tapered capillaries.) Diffracted x-ray or neutrons beams cannot be focused to produce images, so the sample structure must be reconstructed from the diffraction pattern. Sharp features in the diffraction pattern arise from periodic, repeating structure in the sample, which are often very strong due to coherent reflection of many photons from many regularly spaced instances of similar structure, while non-periodic components of the structure result in diffuse (and usually weak) diffraction features. Said more simply, areas with a higher density and repetition of atom order tend to reflect more light toward one point in space when compared to those areas with fewer atoms and less repetition.

Because of their highly ordered and repetitive structure, crystals give diffraction patterns of sharp Bragg reflection spots, and are ideal for analysing the structure of solids.

### Notation

- Coordinates in *square brackets* such as [100] denote a direction vector (in real space).
- Coordinates in *angle brackets* or *chevrons* such as <100> denote a *family* of directions which are related by symmetry operations.

In the cubic crystal system for example, <100> would mean [100], [010], [001] or the negative of any of those directions.

- Miller indices in *parentheses* such as (100) denote a plane of the crystal structure, and regular repetitions of that plane with a particular spacing. In the cubic system, the normal to the (hkl) plane is the direction [hkl], but in lower-symmetry cases, the normal to (hkl) is not parallel to [hkl].
- Indices in *curly brackets* or *braces* such as {100} denote a family of planes and their normals which are equivalent in cubic materials due to symmetry operations, much the way angle brackets denote a family of directions. In non-cubic materials, <hkl> is not necessarily perpendicular to {hkl}.

### *Technique*

Some materials studied using crystallography, proteins for example, do not occur naturally as crystals. Typically, such molecules are placed in solution and allowed to crystallise over days, weeks, or months through vapour diffusion. A drop of solution containing the molecule, buffer, and precipitants is sealed in a container with a reservoir containing a hygroscopic solution. Water in the drop diffuses to the reservoir, slowly increasing the concentration and allowing a crystal to form. If the concentration were to rise more quickly, the molecule would simply precipitate out of solution, resulting in disorderly granules rather than an orderly and hence usable crystal.

Once a crystal is obtained, data can be collected using a beam of radiation. Although many universities that engage in crystallographic research have their own X-ray producing equipment, synchrotrons are often used as X-ray sources, because of the purer and more complete patterns such sources can generate. Synchrotron sources also have a much higher intensity of X-ray beams, so data collection takes a fraction of the time normally necessary at weaker sources. Complementary neutron crystallography techniques are used to enhance hydrogen atoms positions. Such techniques are available in Neutron facilities.

Producing an image from a diffraction pattern requires sophisticated mathematics and often an iterative process of modelling and refinement. In this process, the mathematically predicted diffraction patterns of an hypothesized or “model” structure are compared to the actual pattern generated by the crystalline sample.

Ideally, researchers make several initial guesses, which through refinement all converge on the same answer. Models are refined until their predicted patterns match to as great a degree as can be achieved without radical revision of the model. This is a painstaking process, made much easier today by computers.

The mathematical methods for the analysis of diffraction data only apply to *patterns,* which in turn result only when waves diffract from orderly arrays. Hence crystallography applies for the most part only to crystals, or to molecules which can be coaxed to crystallise for the sake of measurement. In spite of this, a certain amount of molecular information can be deduced from the patterns that are generated by fibers and powders, which while not as perfect as a solid crystal, may exhibit a degree of order. This level of order can be sufficient to deduce the structure of simple molecules, or to determine the coarse features of more complicated molecules. For example, the double-helical structure of DNA was deduced from an X-ray diffraction pattern that had been generated by a fibrous sample.

### Crystallography in Materials Engineering

**Figure 19:** An example of a cubic lattice

Crystallography is a tool that is often employed by materials scientists. In single crystals, the effects of the crystalline arrangement of atoms is often easy to see macroscopically, because the natural shapes of crystals reflect the atomic structure. In addition, physical properties are often controlled by crystalline defects. The understanding of crystal structures is an important prerequisite for understanding crystallographic defects. Mostly, materials do not occur in a single crystalline, but poly-crystalline form, such that the powder diffraction method plays a most important role in structural determination.

A number of other physical properties are linked to crystallography. For example, the minerals in clay form small, flat, platelike structures.

Clay can be easily deformed because the platelike particles can slip along each other in the plane of the plates, yet remain strongly connected in the direction perpendicular to the plates. Such mechanisms can be studied by crystallographic texture measurements. In another example, iron transforms from a body-centered cubic (bcc) structure to a face-centered cubic (fcc) structure called austenite when it is heated. The fcc structure is a close-packed structure, and the bcc structure is not, which explains why the volume of the iron decreases when this transformation occurs.

Crystallography is useful in phase identification. When performing any process on a material, it may be desired to find out what compounds and what phases are present in the material. Each phase has a characteristic arrangement of atoms. Techniques like X-ray or neutron diffraction can be used to identify which patterns are present in the material, and thus which compounds are present. Crystallography covers the enumeration of the symmetry patterns which can be formed by atoms in a crystal and for this reason has a relation to group theory and geometry.

### *Biology*

X-ray crystallography is the primary method for determining the molecular conformations of biological macromolecules, particularly protein and nucleic acids such as DNA and RNA. In fact, the double-helical structure of DNA was deduced from crystallographic data. The first crystal structure of a macromolecule was solved in 1958 A three-dimensional model of the myoglobin molecule obtained by X-ray analysis. The Protein Data Bank (PDB) is a freely accessible repository for the structures of proteins and other biological macromolecules. Computer programs like RasMol or Pymol can be used to visualise biological molecular structures. Neutron crystallography is often used to help refine structures obtained by x-ray methods or to solve a specific bond; the methods are often viewed as complementary, as x-rays are sensitive to electron positions and scatter most strongly off heavy atoms, while neutrons are sensitive to nucleus positions and scatter strongly off many light isotopes, including hydrogen and deuterium. Electron crystallography has been used to determine some protein structures, most notably membrane proteins and viral capsids.

### DNA

Deoxyribonucleic acid is a nucleic acid that contains the genetic instructions used in the development and functioning of all known living organisms (with the exception of RNA viruses). The DNA segments that carry this genetic information are called genes, but other DNA sequences have structural purposes, or are involved in regulating the use of this genetic information. Along with RNA and proteins, DNA is one of the three major macromolecules that are essential for all known forms of life.

DNA consists of two long polymers of simple units called nucleotides, with backbones made of sugars and phosphate groups joined by ester bonds. These two strands run in opposite directions to each other and are therefore anti-parallel. Attached to each sugar is one of four types of molecules called nucleobases (informally, *bases*). It is the sequence of these four nucleobases along the backbone that encodes information. This information is read using the genetic code, which specifies the sequence of the amino acids within proteins. The code is read by copying stretches of DNA into the related nucleic acid RNA in a process called transcription.

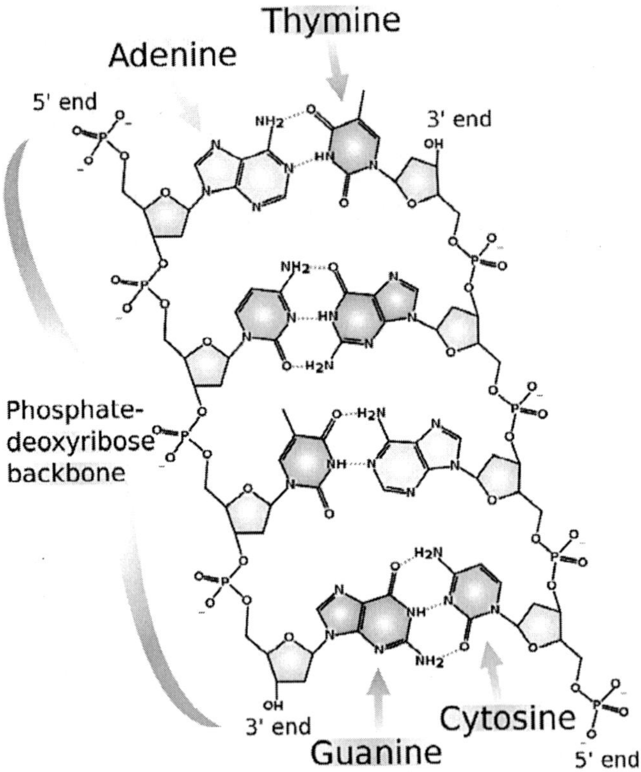


**Figure 20:** Chemical structure of DNA. Hydrogen bonds shown as dotted lines.

Within cells DNA is organised into long structures called chromosomes. During cell division these chromosomes are duplicated in the process of DNA replication, providing each cell its own complete set of chromosomes. Eukaryotic organisms (animals, plants, fungi, and protists) store most of their DNA inside the cell nucleus and some of their DNA in organelles, such as mitochondria or chloroplasts. In contrast, prokaryotes (bacteria and archaea) store their DNA only in the cytoplasm. Within the chromosomes, chromatin proteins such as histones compact and organise DNA. These compact structures guide the interactions between DNA and other proteins, helping control which parts of the DNA are transcribed.

### *Properties*

DNA is a long polymer made from repeating units called nucleotides. As first discovered by James D. Watson and Francis Crick, the structure of DNA of all species comprises two helical chains each coiled round the same axis, and each with a pitch of 34 Engstrϕms (3.4 nanometres) and a radius of 10 Engstrϕms (1.0 nanometres). According to another study, when measured in a particular solution, the DNA chain measured 22 to 26 Engstrϕms wide (2.2 to 2.6 nanometres), and one nucleotide unit measured 3.3 E (0.33 nm) long. Although each individual repeating unit is very small, DNA polymers can be very large molecules containing millions of nucleotides. For instance, the largest human chromosome, chromosome number 1, is approximately 220 million base pairs long.

In living organisms DNA does not usually exist as a single molecule, but instead as a pair of molecules that are held tightly together. These two long strands entwine like vines, in the shape of a double helix. The nucleotide repeats contain both the segment of the backbone of the molecule, which holds the chain together, and a nucleobase, which interacts with the other DNA strand in the helix. A nucleobase linked to a sugar is called a nucleoside and a base linked to a sugar and one or more phosphate groups is called a nucleotide. Polymers comprising multiple linked nucleotides (as in DNA) are called a polynucleotide.

The backbone of the DNA strand is made from alternating phosphate and sugar residues. The sugar in DNA is 2-deoxyribose, which is a pentose (five-carbon) sugar. The sugars are joined together by phosphate groups that form phosphodiester bonds between the third and fifth carbon atoms of adjacent sugar rings. These asymmetric bonds mean a strand of DNA has a direction. In a double helix the direction of the nucleotides in one strand is opposite to their direction in the other strand: the strands are *antiparallel*. The asymmetric ends of DNA strands are called the 52 (*five prime*) and 32 (*three prime*) ends, with the 5' end having a terminal phosphate group and the 3' end a terminal hydroxyl group. One major difference between DNA and RNA is the sugar, with the 2-deoxyribose in DNA being replaced by the alternative pentose sugar ribose in RNA.

The DNA double helix is stabilised primarily by two forces: hydrogen bonds between nucleotides and base-stacking interactions among the aromatic nucleobases. In the aqueous environment of the cell, the conjugated ð bonds of nucleotide bases align perpendicular to the axis of the DNA molecule, minimising their interaction with the solvation shell and therefore, the Gibbs free energy. The four bases found in DNA are adenine (abbreviated A), cytosine (C), guanine (G) and thymine (T). These four bases are attached to the sugar/phosphate to form the complete nucleotide, as shown for adenosine monophosphate.

The nucleobases are classified into two types: the purines, A and G, being fused five- and six-membered heterocyclic compounds, and the pyrimidines, the six-membered rings C and T. A fifth pyrimidine nucleobase, uracil (U), usually takes the place of thymine in RNA and differs from thymine by lacking a methyl group on its ring. Uracil is not usually found in DNA, occurring only as a breakdown product of cytosine. In addition to RNA and DNA a large number of artificial nucleic acid analogues have also been created to study the proprieties of nucleic acids, or for use in biotechnology.

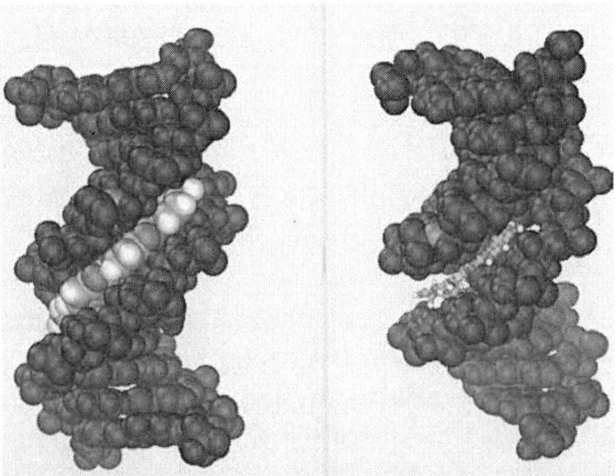

**Figure 21:** Major and minor grooves of DNA. Minor groove is a binding site for the dye Hoechst 33258.

### Grooves

Twin helical strands form the DNA backbone. Another double helix may be found by tracing the spaces, or grooves, between the strands. These voids are adjacent to the base pairs and may provide a binding site. As the strands are not directly opposite each other, the grooves are unequally sized.

One groove, the major groove, is 22 E wide and the other, the minor groove, is 12 E wide. The narrowness of the minor groove means that the edges of the bases are more accessible in the major groove. As a result, proteins like transcription factors that can bind to specific sequences in double-stranded DNA usually make contacts to the sides of the bases exposed in the major groove. This situation varies in unusual conformations of DNA within the cell, but the major and minor grooves are always named to reflect the differences in size that would be seen if the DNA is twisted back into the ordinary B form.

### Base Pairing

In a DNA double helix, each type of nucleobase on one strand normally interacts with just one type of nucleobase on the other strand. This is called complementary base pairing. Here, purines form hydrogen bonds to pyrimidines, with A bonding only to T, and C bonding only to G.

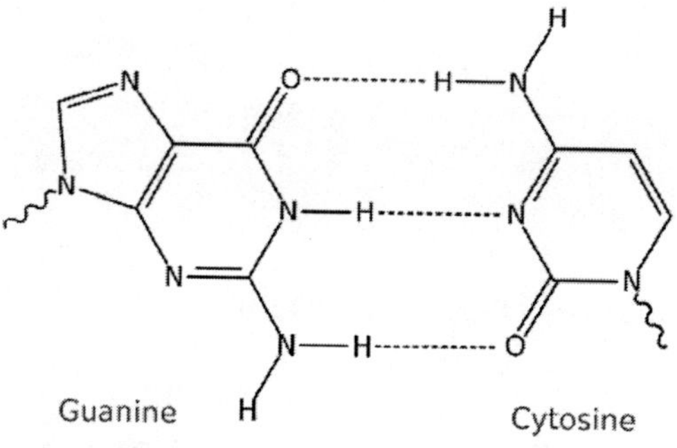


**Figure 22:** Top, a GC base pair with three hydrogen bonds. Bottom, an AT base pair with two hydrogen bonds. Non-covalent hydrogen bonds between the pairs are shown as dashed lines.

This arrangement of two nucleotides binding together across the double helix is called a base pair. As hydrogen bonds are not covalent, they can be broken and rejoined relatively easily. The two strands of DNA in a double helix can therefore be pulled apart like a zipper, either by a mechanical force or high temperature. As a result of this complementarity, all the information in the double-stranded sequence of a DNA helix is duplicated on each strand, which is vital in DNA replication. Indeed, this reversible and specific interaction between complementary base pairs is critical for all the functions of DNA in living organisms.

The two types of base pairs form different numbers of hydrogen bonds, AT forming two hydrogen bonds, and GC forming three hydrogen bonds. DNA with high GC-content is more stable than DNA with low GC-content. Although it is often stated that this is due to the added stability of an additional hydrogen bond, this is incorrect. DNA with high GC-content is more stable due to intra-strand base stacking interactions.

As noted above, most DNA molecules are actually two polymer strands, bound together in a helical fashion by noncovalent bonds; this double stranded structure (dsDNA) is maintained largely by the intrastrand base stacking interactions, which are strongest for G,C stacks. The two strands can come apart – a process known as melting – to form two ss DNA molecules. Melting occurs when conditions favour ssDNA; such conditions are high temperature, low salt and high pH (low pH also melts DNA, but since DNA is unstable due to acid depurination, low pH is rarely used). The stability of the dsDNA form depends not only on the GC-content (% G,C basepairs) but also on sequence (since stacking is sequence specific) and also length (longer molecules are more stable). The stability can be measured in various ways; a common way is the "melting temperature", which is the temperature at which 50% of the ds molecules are converted to ss molecules; melting temperature is dependent on ionic strength and the concentration of DNA. As a result, it is both the percentage of GC base pairs and the overall length of a DNA double helix that determine the strength of the association between the two strands of DNA. Long DNA helices with a high GC-content have stronger-interacting strands, while short helices with high AT content have weaker-interacting strands. In biology, parts of the DNA double helix that need to separate easily, such as the TATAAT Pribnow box in some promoters, tend to have a high AT content, making the strands easier to pull apart.

In the laboratory, the strength of this interaction can be measured by finding the temperature required to break the hydrogen bonds, their melting temperature (also called $T_m$ value). When all the base pairs in a DNA double helix melt, the strands separate and exist in solution as two entirely independent molecules. These single-stranded DNA molecules (*ssDNA*) have no single common shape, but some conformations are more stable than others.

### *Sense and Antisense*

A DNA sequence is called "sense" if its sequence is the same as that of a messenger RNA copy that is translated into protein. The sequence on the opposite strand is called the "antisense" sequence. Both sense and antisense sequences can exist on different parts of the same strand of DNA (i.e. both strands contain both sense and antisense sequences). In both prokaryotes and eukaryotes, antisense RNA sequences are produced, but the functions of these RNAs are not entirely clear. One proposal is that antisense RNAs are involved in regulating gene expression through RNA-RNA base pairing.

A few DNA sequences in prokaryotes and eukaryotes, and more in plasmids and viruses, blur the distinction between sense and antisense strands by having overlapping genes. In these cases, some DNA sequences do double duty, encoding one protein when read along one strand, and a second protein when read in the opposite direction along the other strand. In bacteria, this overlap may be involved in the regulation of gene transcription, while in viruses, overlapping genes increase the amount of information that can be encoded within the small viral genome.

### *Supercoiling*

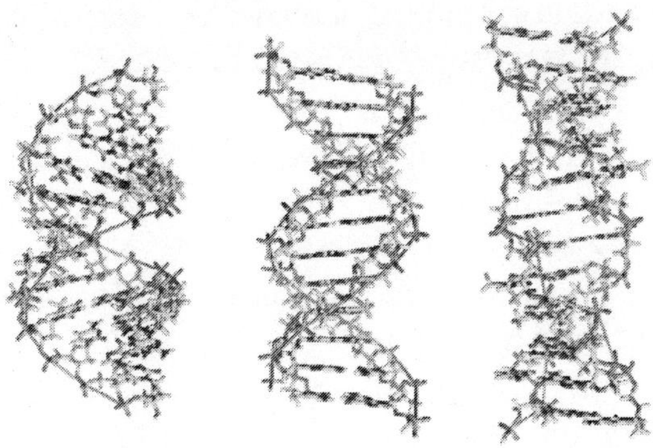

**Figure 23:** From left to right, the structures of A, B and Z DNA

DNA can be twisted like a rope in a process called DNA supercoiling. With DNA in its "relaxed" state, a strand usually circles the axis of the double helix once every 10.4 base pairs, but if the DNA is twisted the strands become more tightly or more loosely wound. If the DNA is twisted in the direction of the helix, this is positive supercoiling, and the bases are held more tightly together. If they are twisted in the opposite direction, this is negative supercoiling, and the bases come apart more easily. In nature, most DNA has slight negative supercoiling that is introduced by enzymes called topoisomerases. These enzymes are also needed to relieve the twisting stresses introduced into DNA strands during processes such as transcription and DNA replication.

### *Alternate DNA Structures*

DNA exists in many possible conformations that include A-DNA, B-DNA, and Z-DNA forms, although, only B-DNA and Z-DNA have been directly observed in functional organisms. The conformation that DNA adopts depends on the hydration level, DNA sequence, the amount and direction of supercoiling, chemical modifications of the bases, the type and concentration of metal ions, as well as the presence of polyamines in solution.

The first published reports of A-DNA X-ray diffraction patterns— and also B-DNA used analyses based on Patterson transforms that provided only a limited amount of structural information for oriented fibers of DNA. An alternate analysis was then proposed by Wilkins *et al.*, in 1953, for the *in vivo* B-DNA X-ray diffraction/scattering patterns of highly hydrated DNA fibers in terms of squares of Bessel functions. In the same journal, James D. Watson and Francis Crick presented their molecular modeling analysis of the DNA X-ray diffraction patterns to suggest that the structure was a double-helix.

Although the 'B-DNA form' is most common under the conditions found in cells, it is not a well-defined conformation but a family of related DNA conformations that occur at the high hydration levels present in living cells. Their corresponding X-ray diffraction and scattering patterns are characteristic of molecular paracrystals with a significant degree of disorder.

Compared to B-DNA, the A-DNA form is a wider right-handed spiral, with a shallow, wide minor groove and a narrower, deeper major groove. The A form occurs under non-physiological conditions in partially dehydrated samples of DNA, while in the cell it may be produced in hybrid pairings of DNA and RNA strands, as well as in enzyme-DNA complexes. Segments of DNA where the bases have been chemically modified by methylation may undergo a larger change in conformation and adopt the Z form. Here, the strands turn about the helical axis in a left-handed spiral, the opposite of the more common B form. These unusual structures can be recognised by specific Z-DNA binding proteins and may be involved in the regulation of transcription.

### *Alternate DNA Chemistry*

For a number of years exobiologists have proposed the existence of a shadow biosphere, a postulated microbial biosphere of Earth that uses radically different biochemical and molecular processes than currently known life. One of the proposals was the existence of lifeforms that use arsenic instead of phosphorus in DNA.

A December 2010 NASA press conference stated that the bacterium GFAJ-1, which has evolved in an arsenic-rich environment, is the first terrestrial lifeform found which may have this ability. The bacterium was found in Mono Lake, east of Yosemite National Park.

GFAJ-1 is a rod-shaped extremophile bacterium in the family Halomonadaceae that, when starved of phosphorus, may be capable of incorporating the usually poisonous element arsenic in its DNA. This discovery may lend weight to the long-standing idea that extraterrestrial life could have a different chemical makeup from life on Earth. The research was carried out by a team led by Felisa Wolfe-Simon, a geomicrobiologist and geobiochemist, a Postdoctoral Fellow of the NASA Astrobiology Institute with Arizona State University. This finding has, however, faced strong criticism from the scientific community; scientists have argued that there is no evidence that arsenic is actually incorporated into biomolecules. Independent conformation of this finding has also not yet been possible.

### *Quadruplex Structures*

At the ends of the linear chromosomes are specialised regions of DNA called telomeres. The main function of these regions is to allow the cell to replicate chromosome ends using the enzyme telomerase, as the enzymes that normally replicate DNA cannot copy the extreme 32 ends of chromosomes. These specialised chromosome caps also help protect the DNA ends, and stop the DNA repair systems in the cell from treating them as damage to be corrected. Inhuman cells, telomeres are usually lengths of single-stranded DNA containing several thousand repeats of a simple TTAGGG sequence.

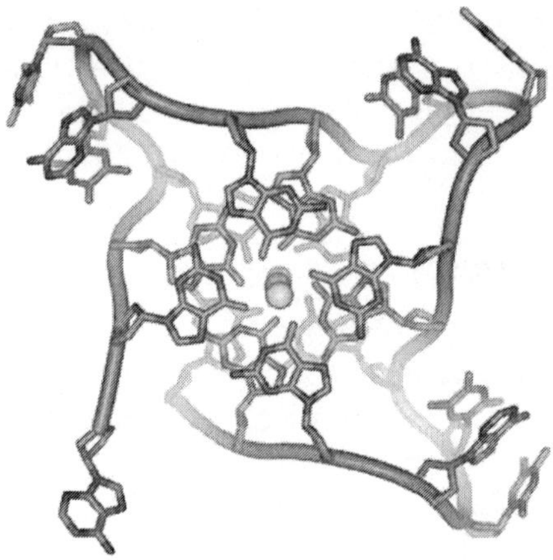

**Figure 24:** DNA quadruplex formed by telomererepeats. The looped conformation of the DNA backbone is very different from the typical DNA helix.

These guanine-rich sequences may stabilise chromosome ends by forming structures of stacked sets of four-base units, rather than the usual base pairs found in other DNA molecules. Here, four guanine bases form a flat plate and these flat four-base units then stack on top of each other, to form a stable *G-quadruplex* structure. These structures are stabilised by hydrogen bonding between the edges of the bases and chelation of a metal ion in the centre of each four-base unit. Other structures can also be formed, with the central set of four bases coming from either a single strand folded around the bases, or several different parallel strands, each contributing one base to the central structure.

In addition to these stacked structures, telomeres also form large loop structures called telomere loops, or T-loops. Here, the single-stranded DNA curls around in a long circle stabilised by telomere-binding proteins. At the very end of the T-loop, the single-stranded telomere DNA is held onto a region of double-stranded DNA by the telomere strand disrupting the double-helical DNA and base pairing to one of the two strands. This triple-stranded structure is called a displacement loop or D-loop.

### Branched DNA

In DNA fraying occurs when non-complementary regions exist at the end of an otherwise complementary double-strand of DNA. However, branched DNA can occur if a third strand of DNA is introduced and contains adjoining regions able to hybridise with the frayed regions of the pre-existing double-strand. Although the simplest example of branched DNA involves only three strands of DNA, complexes involving additional strands and multiple branches are also possible.

### Vibration

DNA may carry out low-frequency collective motion as observed by the Raman spectroscopy and analysed with a quasi-continuum model.

### Chemical Modifications

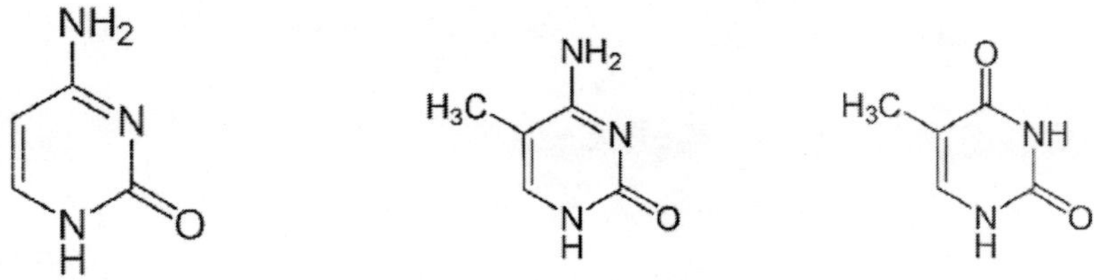


cytosine 5-methylcytosine thymine

**Figure 25:** Structure of cytosine with and without the 5-methyl group. Deamination converts 5-methylcytosine into thymine.

### Base Modifications

The expression of genes is influenced by how the DNA is packaged in chromosomes, in a structure called chromatin. Base modifications can be involved in packaging, with regions that have low or no gene expression usually containing high levels of methylation of cytosine bases.

For example, cytosine methylation, produces 5-methylcytosine, which is important for X-chromosome inactivation. The average level of methylation varies between organisms – the worm *Caenorhabditis elegans* lacks cytosine methylation, while vertebrates have higher levels, with up to 1% of their DNA containing 5-methylcytosine.

Despite the importance of 5-methylcytosine, it can deaminate to leave a thymine base, so methylated cytosines are particularly prone to mutations.

Other base modifications include adenine methylation in bacteria, the presence of 5-hydroxymethylcytosine in the brain, and the glycosylation of uracil to produce the "J-base" inkinetoplastids.

### Damage

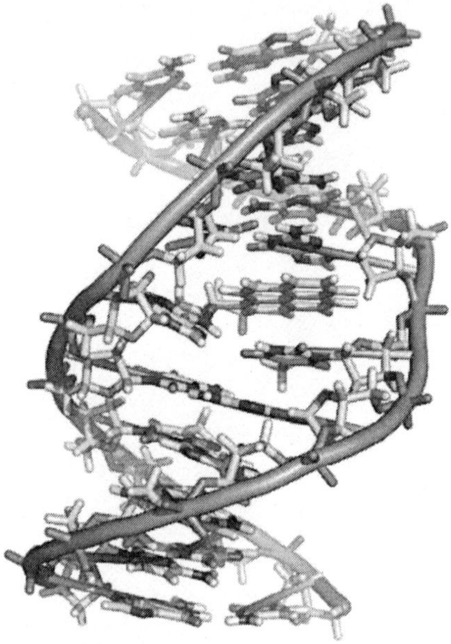

**Figure 26:** A covalent adduct between a metabolically activated form of benzo[*a*]pyrene, the major mutagen in tobacco smoke, and DNA

DNA can be damaged by many sorts of mutagens, which change the DNA sequence. Mutagens include oxidising agents, alkylating agents and also high-energy electromagnetic radiation such as ultraviolet light and X-rays. The type of DNA damage produced depends on the type of mutagen. For example, UV light can damage DNA by producing thymine dimers, which are cross-links between pyrimidine bases. On the other hand, oxidants such as free radicals or hydrogen peroxide produce multiple forms of damage, including base modifications, particularly of guanosine, and double-strand breaks. A typical human cell contains about 150,000 bases that have suffered oxidative damage. Of these oxidative lesions, the most dangerous are double-strand breaks, as these are difficult to repair and can produce point mutations, insertions and deletions from the DNA sequence, as well as chromosomal translocations.

Many mutagens fit into the space between two adjacent base pairs, this is called *intercalation*. Most intercalators are aromatic and planar molecules; examples include ethidium bromide, acridines, daunomycin, and doxorubicin. In order for an intercalator to fit between base pairs, the bases must separate, distorting the DNA strands by unwinding of the double helix. This inhibits both transcription and DNA replication, causing toxicity and mutations. As a result, DNA intercalators may be carcinogens, and in the case of thalidomide, a teratogen. Others such as benzo[*a*]pyrene diol epoxide and aflatoxin form DNA adducts which induce errors in replication. Nevertheless, due to their ability to inhibit DNA transcription and replication, other similar toxins are also used in chemotherapy to inhibit rapidly growing cancer cells.

### Biological Functions

DNA usually occurs as linear chromosomes in eukaryotes, and circular chromosomes in prokaryotes. The set of chromosomes in a cell makes up its genome; the human genome has approximately 3 billion base pairs of DNA arranged into 46 chromosomes. The information carried by DNA is held in the sequence of pieces of DNA called genes. Transmission of genetic information in genes is achieved via complementary base pairing.

For example, in transcription, when a cell uses the information in a gene, the DNA sequence is copied into a complementary RNA sequence through the attraction between the DNA and the correct RNA nucleotides. Usually, this RNA copy is then used to make a matching protein sequence in a process called translation, which depends on the same interaction between RNA nucleotides. In alternative fashion, a cell may simply copy its genetic information in a process called DNA replication. The details of these functions are covered in other articles; here we focus on the interactions between DNA and other molecules that mediate the function of the genome.

### Genes and Genomes

Genomic DNA is tightly and orderly packed in the process called DNA condensation to fit the small available volumes of the cell. In eukaryotes, DNA is located in the cell nucleus, as well as small amounts in mitochondria and chloroplasts. In prokaryotes, the DNA is held within an irregularly shaped body in the cytoplasm called the nucleoid. The genetic information in a genome is held within genes, and the complete set of this information in an organism is called its genotype. A gene is a unit of heredity and is a region of DNA that influences a particular characteristic in an organism. Genes contain an open reading frame that can be transcribed, as well as regulatory sequences such as promoters and enhancers, which control the transcription of the open reading frame. In many species, only a small fraction of the total sequence of the genome encodes protein. For example, only about 1.5% of the human genome consists of protein-coding exons, with over 50% of human DNA consisting of non-coding repetitive sequences.

The reasons for the presence of so much noncoding DNA in eukaryotic genomes and the extraordinary differences in genome size, or *C-value*, among species represent a long-standing puzzle known as the "C-value enigma". However, DNA sequences that do not code protein may still encode functional non-coding RNA molecules, which are involved in the regulation of gene expression.

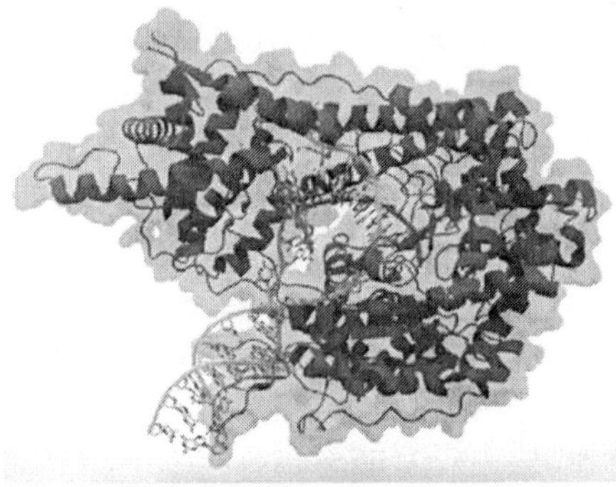

**Figure 27:** T7 RNA polymerase producing a mRNA from a DNA template.

Some noncoding DNA sequences play structural roles in chromosomes. Telomeres and centromeres typically contain few genes, but are important for the function and stability of chromosomes. An abundant form of noncoding DNA in humans are pseudogenes, which are copies of genes that have been disabled by mutation. These sequences are usually just molecular fossils, although they can occasionally serve as raw genetic material for the creation of new genes through the process of gene duplication and divergence.

### *Transcription and Translation*

A gene is a sequence of DNA that contains genetic information and can influence the phenotype of an organism. Within a gene, the sequence of bases along a DNA strand defines a messenger RNAsequence, which then defines one or more protein sequences. The relationship between the nucleotide sequences of genes and the amino-acid sequences of proteins is determined by the rules of translation, known collectively as the genetic code. The genetic code consists of three-letter 'words' called *codons* formed from a sequence of three nucleotides (e.g. ACT, CAG, TTT).

In transcription, the codons of a gene are copied into messenger RNA by RNA polymerase. This RNA copy is then decoded by a ribosome that reads the RNA sequence by base-pairing the messenger RNA to transfer RNA, which carries amino acids. Since there are 4 bases in 3-letter combinations, there are 64 possible codons ($4^3$ combinations). These encode the twenty standard amino acids, giving most amino acids more than one possible codon. There are also three 'stop' or 'nonsense' codons signifying the end of the coding region; these are the TAA, TGA and TAG codons.

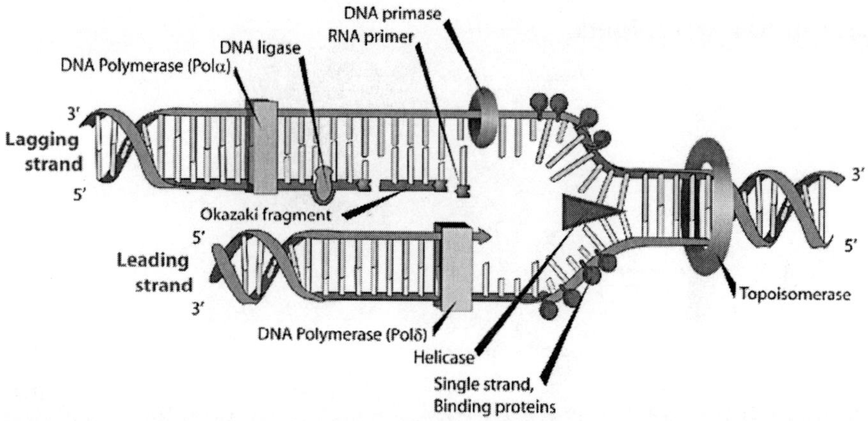


**Figure 28:** DNA replication. The double helix is unwound by a helicase and topoisomerase. Next, one DNA polymerase produces the leading strand copy. Another DNA polymerase binds to the lagging strand. This enzyme makes discontinuous segments (called Okazaki fragments) before DNA ligase joins them together.

### *Replication*

Cell division is essential for an organism to grow, but, when a cell divides, it must replicate the DNA in its genome so that the two daughter cells have the same genetic information as their parent. The double-stranded structure of DNA provides a simple mechanism for DNA replication. Here, the two strands are separated and then each strand's complementary DNA sequence is recreated by an enzyme called DNA polymerase. This enzyme makes the complementary strand by finding the correct base through complementary base pairing, and bonding it onto the original strand.

As DNA polymerases can only extend a DNA strand in a 52 to 32 direction, different mechanisms are used to copy the antiparallel strands of the double helix. In this way, the base on the old strand dictates which base appears on the new strand, and the cell ends up with a perfect copy of its DNA.

### Interactions with Proteins

All the functions of DNA depend on interactions with proteins. These protein interactions can be non-specific, or the protein can bind specifically to a single DNA sequence. Enzymes can also bind to DNA and of these, the polymerases that copy the DNA base sequence in transcription and DNA replication are particularly important.

### DNA-binding Proteins

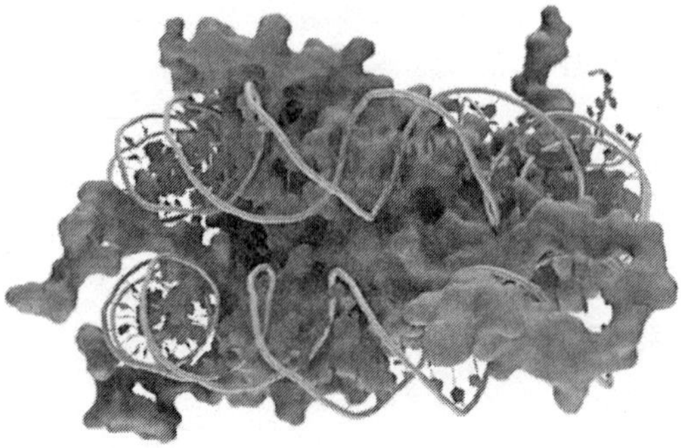

**Figure 29:** Interaction of DNA with histones. These proteins' basic amino acids bind to the acidic phosphate groups on DNA.

Structural proteins that bind DNA are well-understood examples of non-specific DNA-protein interactions. Within chromosomes, DNA is held in complexes with structural proteins. These proteins organise the DNA into a compact structure called chromatin. In eukaryotes this structure involves DNA binding to a complex of small basic proteins called histones, while in prokaryotes multiple types of proteins are involved. The histones form a disk-shaped complex called a nucleosome, which contains two complete turns of double-stranded DNA wrapped around its surface. These non-specific interactions are formed through basic residues in the histones making ionic bonds to the acidic sugar-phosphate backbone of the DNA, and are therefore largely independent of the base sequence. Chemical modifications of these basic amino acid residues include methylation, phosphorylation and acetylation. These chemical changes alter the strength of the interaction between the DNA and the histones, making the DNA more or less accessible to transcription factors and changing the rate of transcription. Other non-specific DNA-binding proteins in chromatin include the high-mobility group proteins, which bind to bent or distorted DNA. These proteins are important in bending arrays of nucleosomes and arranging them into the larger structures that make up chromosomes.

A distinct group of DNA-binding proteins are the DNA-binding proteins that specifically bind single-stranded DNA. In humans, replication protein A is the best-understood member of this family and is used in processes where the double helix is separated, including DNA replication, recombination and DNA repair. These binding proteins seem to stabilise single-stranded DNA and protect it from forming stem-loops or being degraded by nucleases.

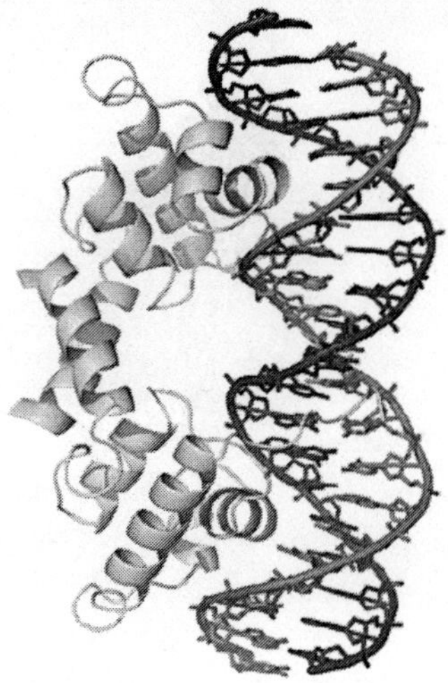

**Figure 30:** The lambda repressor helix-turn-helix transcription factor bound to its DNA target

In contrast, other proteins have evolved to bind to particular DNA sequences. The most intensively studied of these are the various transcription factors, which are proteins that regulate transcription. Each transcription factor binds to one particular set of DNA sequences and activates or inhibits the transcription of genes that have these sequences close to their promoters. The transcription factors do this in two ways. Firstly, they can bind the RNA polymerase responsible for transcription, either directly or through other mediator proteins; this locates the polymerase at the promoter and allows it to begin transcription. Alternatively, transcription factors can bindenzymes that modify the histones at the promoter; this will change the accessibility of the DNA template to the polymerase.

As these DNA targets can occur throughout an organism's genome, changes in the activity of one type of transcription factor can affect thousands of genes. Consequently, these proteins are often the targets of the signal transduction processes that control responses to environmental changes or cellular differentiation and development. The specificity of these transcription factors' interactions with DNA come from the proteins making multiple contacts to the edges of the DNA bases, allowing them to "read" the DNA sequence. Most of these base-interactions are made in the major groove, where the bases are most accessible.

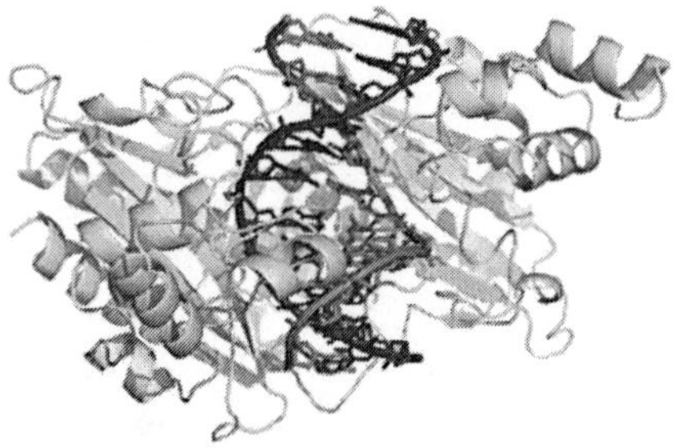

**Figure 31:** The restriction enzyme EcoRV in a complex with its substrate DNA

### DNA-modifying Enzymes

***Nucleases and Ligases:*** Nucleases are enzymes that cut DNA strands by catalysing the hydrolysis of the phosphodiester bonds. Nucleases that hydrolyse nucleotides from the ends of DNA strands are called exonucleases, while endonucleases cut within strands. The most frequently used nucleases in molecular biology are the restriction endonucleases, which cut DNA at specific sequences. For instance, the EcoRV enzyme shown to the left recognises the 6-base sequence 52 -GAT | ATC-32 and makes a cut at the vertical line. In nature, these enzymes protect bacteria against phage infection by digesting the phage DNA when it enters the bacterial cell, acting as part of the restriction modification system. In technology, these sequence-specific nucleases are used in molecular cloning and DNA fingerprinting.

Enzymes called DNA ligases can rejoin cut or broken DNA strands. Ligases are particularly important in lagging strand DNA replication, as they join together the short segments of DNA produced at the replication fork into a complete copy of the DNA template. They are also used in DNA repair and genetic recombination.

### *Topoisomerases and Helicases*

Topoisomerases are enzymes with both nuclease and ligase activity. These proteins change the amount of supercoiling in DNA. Some of these enzymes work by cutting the DNA helix and allowing one section to rotate, thereby reducing its level of supercoiling; the enzyme then seals the DNA break. Other types of these enzymes are capable of cutting one DNA helix and then passing a second strand of DNA through this break, before rejoining the helix. Topoisomerases are required for many processes involving DNA, such as DNA replication and transcription.

Helicases are proteins that are a type of molecular motor. They use the chemical energy in nucleoside triphosphates, predominantly ATP, to break hydrogen bonds between bases and unwind the DNA double helix into single strands. These enzymes are essential for most processes where enzymes need to access the DNA bases.

### *Polymerases*

Polymerases are enzymes that synthesize polynucleotide chains from nucleoside triphosphates. The sequence of their products are copies of existing polynucleotide chains – which are called *templates*. These enzymes function by adding nucleotides onto the 32 hydroxyl group of the previous nucleotide in a DNA strand. As a consequence, all polymerases work in a 52 to 32 direction. In the active site of these enzymes, the incoming nucleoside triphosphate base-pairs to the template: this allows polymerases to accurately synthesize the complementary strand of their template. Polymerases are classified according to the type of template that they use.

In DNA replication, a DNA-dependent DNA polymerase makes a copy of a DNA sequence. Accuracy is vital in this process, so many of these polymerases have a proofreading activity. Here, the polymerase recognises the occasional mistakes in the synthesis reaction by the lack of base pairing between the mismatched nucleotides. If a mismatch is detected, a 32 to 52 exonuclease activity is activated and the incorrect base removed. In most organisms, DNA polymerases function in a large complex called the replisome that contains multiple accessory subunits, such as the DNA clamp or helicases.

RNA-dependent DNA polymerases are a specialised class of polymerases that copy the sequence of an RNA strand into DNA. They include reverse transcriptase, which is a viral enzyme involved in the infection of cells by retroviruses, and telomerase, which is required for the replication of telomeres. Telomerase is an unusual polymerase because it contains its own RNA template as part of its structure.

Transcription is carried out by a DNA-dependent RNA polymerase that copies the sequence of a DNA strand into RNA. To begin transcribing a gene, the RNA polymerase binds to a sequence of DNA called a promoter and separates the DNA strands.

It then copies the gene sequence into a messenger RNA transcript until it reaches a region of DNA called the terminator, where it halts and detaches from the DNA. As with human DNA-dependent DNA polymerases, RNA polymerase II, the enzyme that transcribes most of the genes in the human genome, operates as part of a large protein complex with multiple regulatory and accessory subunits.

### *Genetic Recombination*

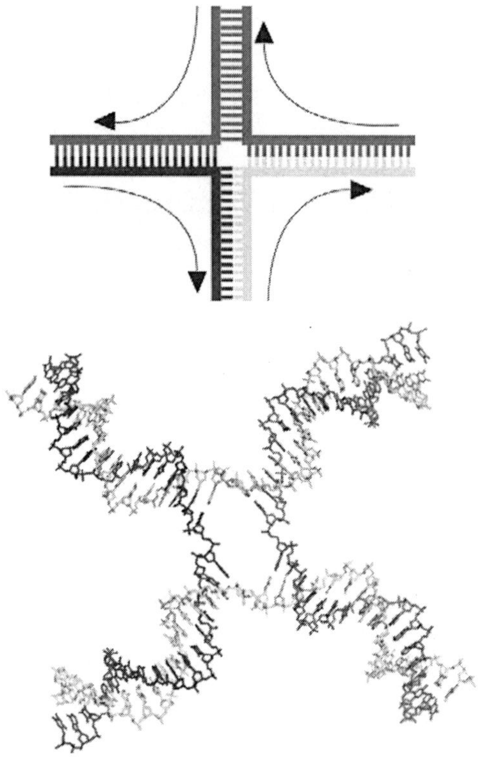

**Figure 32:** Structure of the Holliday junction intermediate in genetic recombination. The four separate DNA strands are coloured red, blue, green and yellow.

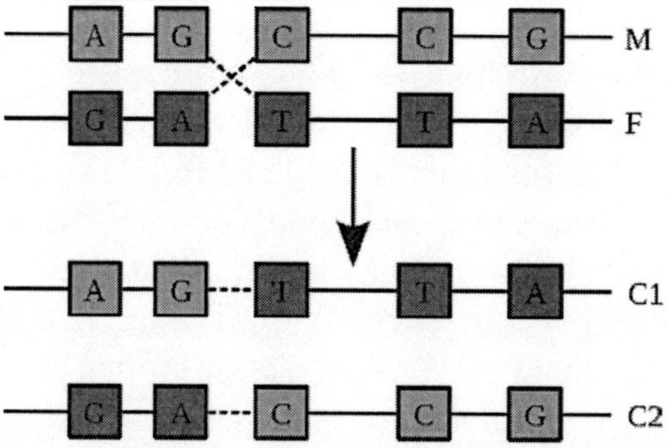


**Figure 33:** Recombination involves the breakage and rejoining of two chromosomes (M and F) to produce two rearranged chromosomes (C1 and C2).

A DNA helix usually does not interact with other segments of DNA, and in human cells the different chromosomes even occupy separate areas in the nucleus called "chromosome territories". This physical separation of different chromosomes is important for the ability of DNA to function as a stable repository for information, as one of the few times chromosomes interact is during chromosomal crossover when they recombine. Chromosomal crossover is when two DNA helices break, swap a section and then rejoin.

Recombination allows chromosomes to exchange genetic information and produces new combinations of genes, which increases the efficiency of natural selection and can be important in the rapid evolution of new proteins. Genetic recombination can also be involved in DNA repair, particularly in the cell's response to double-strand breaks.

The most common form of chromosomal crossover is homologous recombination, where the two chromosomes involved share very similar sequences. Non-homologous recombination can be damaging to cells, as it can produce chromosomal translocations and genetic abnormalities. The recombination reaction is catalysed by enzymes known as recombinases, such as RAD51. The first step in recombination is a double-stranded break either caused by an endonuclease or damage to the DNA. A series of steps catalysed in part by the recombinase then leads to joining of the two helices by at least one Holliday junction, in which a segment of a single strand in each helix is annealed to the complementary strand in the other helix. The Holliday junction is a tetrahedral junction structure that can be moved along the pair of chromosomes, swapping one strand for another. The recombination reaction is then halted by cleavage of the junction and religation of the released DNA.

### Evolution

DNA contains the genetic information that allows all modern living things to function, grow and reproduce. However, it is unclear how long in the 4-billion-year history of life DNA has performed this function, as it has been proposed that the earliest forms of life may have used RNA as their genetic material. RNA may have acted as the central part of early cell metabolism as it can both transmit genetic information and carry out catalysis as part of ribozymes. This ancient RNA world where nucleic acid would have been used for both catalysis and genetics may have influenced the evolution of the current genetic code based on four nucleotide bases. This would occur, since the number of different bases in such an organism is a trade-off between a small number of bases increasing replication accuracy and a large number of bases increasing the catalytic efficiency of ribozymes.

However, there is no direct evidence of ancient genetic systems, as recovery of DNA from most fossils is impossible. This is because DNA will survive in the environment for less than one million years and slowly degrades into short fragments in solution. Claims for older DNA have been made, most notably a report of the isolation of a viable bacterium from a salt crystal 250 million years old, but these claims are controversial.

On August 8, 2011, a report, based on NASA studies with meteorites found on Earth, was published suggesting building blocks of DNA (adenine, guanine and related organic molecules) may have been formed extra terrestrially in outer space.

### Uses in Technology

***Genetic Engineering:*** Methods have been developed to purify DNA from organisms, such as phenol-chloroform extraction, and to manipulate it in the laboratory, such as restriction digests and the polymerase chain reaction. Modern biology and biochemistry make intensive use of these techniques in recombinant DNA technology. Recombinant DNA is a man-made DNA sequence that has been assembled from other DNA sequences. They can be transformed into organisms in the form of plasmids or in the appropriate format, by using a viral vector. The genetically modified organisms produced can be used to produce products such as recombinant proteins, used in medical research, or be grown in agriculture.

### Forensics

Forensic scientists can use DNA in blood, semen, skin, saliva or hair found at a crime scene to identify a matching DNA of an individual, such as a perpetrator. This process is formally termed DNA profiling, but may also be called “genetic fingerprinting”. In DNA profiling, the lengths of variable sections of repetitive DNA, such as short tandem repeats and minisatellites, are compared between people. This method is usually an extremely reliable technique for identifying a matching DNA. However, identification can be complicated if the scene is contaminated with DNA from several people. DNA profiling was developed in 1984 by British geneticist Sir Alec Jeffreys, and first used in forensic science to convict Colin Pitchfork in the 1988 Enderby murders case.

People convicted of certain types of crimes may be required to provide a sample of DNA for a database. This has helped investigators solve old cases where only a DNA sample was obtained from the scene. DNA profiling can also be used to identify victims of mass casualty incidents. On the other hand, many convicted people have been released from prison on the basis of DNA techniques, which were not available when a crime had originally been committed.

### Bioinformatics

Bioinformatics involves the manipulation, searching, and data mining of biological data, and this includes DNA sequence data. The development of techniques to store and search DNA sequences have led to widely applied advances in computer science, especially string searching algorithms, machine learning and database theory. String searching or matching algorithms, which find an occurrence of a sequence of letters inside a larger sequence of letters, were developed to search for specific sequences of nucleotides. The DNA sequenced may be aligned with other DNA sequences to identify homologous sequences and locate the specific mutations that make them distinct. These techniques, especially multiple sequence alignment, are used in studying phylogenetic relationships and protein function. Data sets representing entire genomes’ worth of DNA sequences, such as those produced by the Human Genome Project, are difficult to use without the annotations that identify the locations of genes and regulatory elements on each chromosome. Regions of DNA sequence that have the characteristic patterns associated with protein- or RNA-coding genes can be identified by gene finding algorithms, which allow researchers to predict the presence of particular gene products and their possible functions in an organism even before they have been isolated experimentally. Entire genomes may also be compared which can shed light on the evolutionary history of particular organism and permit the examination of complex evolutionary events.

### DNA Nanotechnology

DNA nanotechnology uses the unique molecular recognition properties of DNA and other nucleic acids to create self-assembling branched DNA complexes with useful properties. DNA is thus used as a structural material rather than as a carrier of biological information. This has led to the creation of two-dimensional periodic lattices (both tile-based as well as using the "DNA origami" method) as well as three-dimensional structures in the shapes of polyhedra. Nanomechanical devices and algorithmic self-assembly have also been demonstrated, and these DNA structures have been used to template the arrangement of other molecules such as gold nanoparticles and streptavidin proteins.

### History and Anthropology

Because DNA collects mutations over time, which are then inherited, it contains historical information, and, by comparing DNA sequences, geneticists can infer the evolutionary history of organisms, their phylogeny. This field of phylogenetics is a powerful tool in evolutionary biology. If DNA sequences within a species are compared, population geneticists can learn the history of particular populations. This can be used in studies ranging from ecological genetics to anthropology; For example, DNA evidence is being used to try to identify the Ten Lost Tribes of Israel.

DNA has also been used to look at modern family relationships, such as establishing family relationships between the descendants of Sally Hemings and Thomas Jefferson. This usage is closely related to the use of DNA in criminal investigations detailed above. Indeed, some criminal investigations have been solved when DNA from crime scenes has matched relatives of the guilty individual.

### History of DNA Research

DNA was first isolated by the Swiss physician Friedrich Miescher who, in 1869, discovered a microscopic substance in the pus of discarded surgical bandages. As it resided in the nuclei of cells, he called it "nuclein". In 1878, Albrecht Kossel isolated the non-protein component of "nuclein", nucleic acid, and later isolated its five primary nucleobases. In 1919, Phoebus Levene identified the base, sugar and phosphate nucleotide unit. Levene suggested that DNA consisted of a string of nucleotide units linked together through the phosphate groups. However, Levene thought the chain was short and the bases repeated in a fixed order. In 1937 William Astbury produced the first X-ray diffraction patterns that showed that DNA had a regular structure.

In 1927 Nikolai Koltsov proposed that inherited traits would be inherited via a "giant hereditary molecule" which would be made up of "two mirror strands that would replicate in a semi-conservative fashion using each strand as a template". In 1928, Frederick Griffith discovered that traits of the "smooth" form of the *Pneumococcus* could be transferred to the "rough" form of the same bacteria by mixing killed "smooth" bacteria with the live "rough" form. This system provided the first clear suggestion that DNA carries genetic information—the Avery–MacLeod–McCarty experiment—when Oswald Avery, along with coworkers Colin MacLeod and Maclyn McCarty, identified DNA as the transforming principle in 1943. DNA's role in heredity was confirmed in 1952, when Alfred Hershey andMartha Chase in the Hershey–Chase experiment showed that DNA is the genetic material of the T2 phage.

In 1953, James D. Watson and Francis Crick suggested what is now accepted as the first correct double-helix model of DNA structure in the journal *Nature.* Their double-helix, molecular model of DNA was then based on a single X-ray diffraction image (labelled as "Photo 51") taken by Rosalind Franklin and Raymond Gosling in May 1952, as well as the information that the DNA bases are paired — also obtained through private communications from Erwin Chargaff in the previous years. Chargaff's rules played a very important role in establishing double-helix configurations for B-DNA as well as A-DNA.

Experimental evidence supporting the Watson and Crick model were published in a series of five articles in the same issue of *Nature.* Of these, Franklin and Gosling's paper was the first publication of their own X-ray diffraction data and original analysis method that partially supported the Watson and Crick model; this issue also contained an article on DNA structure by Maurice Wilkins and two of his colleagues, whose analysis and *in vivo* B-DNA X-ray patterns also supported the presence *in vivo* of the double-helical DNA configurations as proposed by Crick and Watson for their double-helix molecular model of DNA in the previous two pages of *Nature*. In 1962, after Franklin's death, Watson, Crick, and Wilkins jointly received the Nobel Prize in Physiology or Medicine. However, Nobel rules of the time allowed only living recipients, but a vigorous debate continues on who should receive credit for the discovery.

In an influential presentation in 1957, Crick laid out the central dogma of molecular biology, which foretold the relationship between DNA, RNA, and proteins, and articulated the "adaptor hypothesis". Final confirmation of the replication mechanism that was implied by the double-helical structure followed in 1958 through the Meselson–Stahl experiment. Further work by Crick and coworkers showed that the genetic code was based on non-overlapping triplets of bases, called codons, allowing Har Gobind Khorana, Robert W. Holley and Marshall Warren Nirenberg to decipher the genetic code. These findings represent the birth of molecular biology.

# Chapter 6

# Photochemistry

Photochemistry, a sub-discipline of chemistry, is the study of chemical reactions that proceed with the absorption of light by atoms or molecules.. Everyday examples include photosynthesis, the degradation of plastics and the formation of vitamin D with sunlight.

## Principles

Light is a type of electromagnetic radiation, a source of energy. The Grotthuss–Draper law (for chemists Theodor Grotthuss and John W. Draper), states that light must be absorbed by a chemical substance in order for a photochemical reaction to take place. For each photon of light absorbed by a chemical system, no more than one molecule is activated for a photochemical reaction, as defined by the quantum yield.

Chemical reactions occur only when a molecule is provided the necessary "activation energy". A simple example can be the combustion of gasoline (a hydrocarbon) into carbon dioxide and water. In this reaction, space they occupy. This is necessarily an icon: in the quantum mechanical representation of molecules, there are only (positively charged) nuclei and a "cloud" of negative electrons. The electron cloud defines an approximate size for the molecule, though there can be no single precise definition of size. For many years the size of atoms has been approximated by mechanical models (CPK), where the atoms have been represented by plastic spheres whose radius (van der Waals radius) describes a sphere within which "most" of the electron density at low pressure, this enables scientists to observe the energy distribution of the products of a chemical reaction before the differences in energy have been smeared out and averaged by repeated collisions.

The absorption of a photon of light by a reactant molecule may also permit a reaction to occur not just by bringing the molecule to the necessary activation energy, but also by changing the symmetry of the molecule's electronic configuration, enabling an otherwise inaccessible reaction path, as described by the Woodward-Hoffmann selection rules. A 2+2 cycloaddition reaction is one example of a pericyclic reaction that can be analysed using these rules or by the related frontier molecular orbital theory.

Photochemical reactions involve electronic reorganisation initiated by electromagnetic radiation. The reactions are several orders of magnitude faster than thermal reactions; reactions as fast as $10^{''9}$ seconds and associated processes as fast as $10^{''15}$ seconds are often observed.

## Spectral Regions

Photochemists typically work in only a few sections of the electromagnetic spectrum. Some of the most widely used sections, and their wavelengths, are the following:

- Ultraviolet: 100–400 nm
- Visible Light: 400–700 nm
- Near infrared: 700–2500 nm

## Applications

Many important processes involve photochemistry. The premier example is photosynthesis, in which most plants use solar energy to convert carbon dioxide and water into glucose, disposing of oxygen as a side-product. Humans rely on photochemistry for the formation of vitamin D. In fireflies, an enzyme in the abdomen catalyses a reaction that results in bioluminescence.

Photochemistry can also be highly destructive. Medicine bottles are often made with darkened glass to prevent the drugs from photodegradation. A pervasive reaction is the generation of singlet oxygen by photosensitized reactions of triplet oxygen. Typical photosensitizers include tetraphenylporphyrin and methylene blue. The resulting singlet oxygen is an aggressive oxidant, capable of converting C-H bonds into C-OH groups. In photodynamic therapy, light is used to destroy tumours by the action of singlet oxygen. Many polymerisations are started by photoinitiatiors, which decompose upon absorbing light to produce the free radicals for Radical polymerisation.

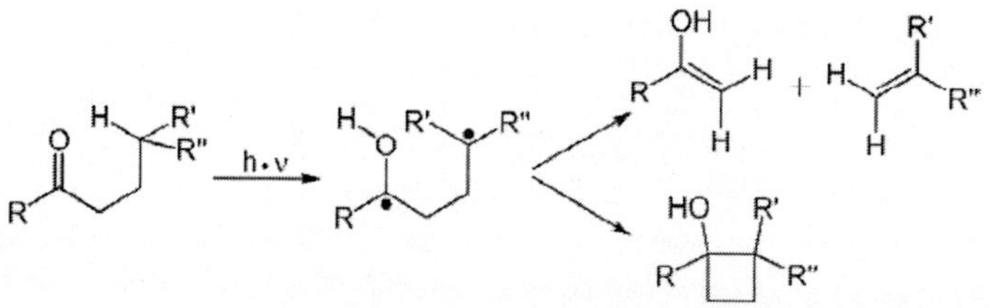


In the area of photochemistry, a photochemical reaction is a chemical reaction that is induced by light. Photochemical reactions are valuable in organic and inorganic chemistry because they proceed differently than thermal reactions. Photochemical reactions are not only very useful but also can be a serious nuisance, as in the photodegradation of many materials, e.g. polyvinyl chloride. A large-scale application of photochemistry is photoresist technology, used in the production of microelectronic components. Vision is initiated by a photochemical reaction of rhodopsin.

### *Experimental Set-up*

Photochemical reactions require a light source that emits wavelengths corresponding to an electronic transition in the reactant. In the early experiments (and in everyday life), sunlight was the light source, although it is polychromatic. Mercury-vapor lamps are more common in the laboratory. Low pressure mercury vapor lamps mainly emit at 254 nm. For polychromatic sources, wavelength ranges can be selected using filters. Alternatively, LEDs and Rayonet lamps emit monochromatically.

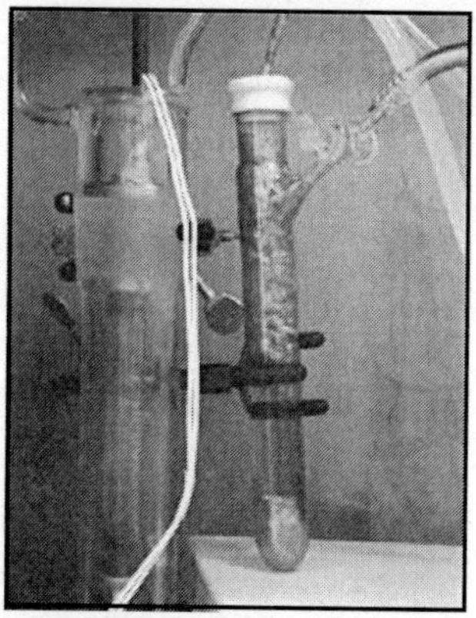

**Figure 1:** Schlenk tube containing slurry of orange crystals of $Fe_2(CO)_9$ in acetic acid after its photochemical synthesis from $Fe(CO)_5$. Themercury lamp (connected to white power cords) can be seen on the left, set inside a water-jacketed quartz tube.

The emitted light must of course reach the targeted functional group without being blocked by the reactor, medium, or other functional groups present. For many applications, quartz is used for the reactors as well as to contain the lamp. Pyrex absorbs at wavelengths shorter than 275 nm. The solvent is an important experimental parameter. Solvents are potential reactants and for this reason, chlorinated solvents are avoided because the C-Cl bond can lead to chlorination of the substrate. Strongly absorbing solvents prevent photons from reaching the substrate. Hydrocarbon solvents absorb only at short wavelengths and are thus preferred for photochemical experiments requiring high energy photons. Solvents containing unsaturation absorb at longer wavelengths and can usefully filter out short wavelengths. For example, cyclohexane and acetone "cut off" (absorb strongly) at wavelengths shorter than 215 and 330 nm, respectively.

## Excitation

Photoexcitation is the first step in a photochemical process where the reactant is elevated to a state of higher energy, an excited state. The photon can be absorbed directly by the reactant or by a photosensitizers, which absorbs the photon and transfers the energy to the reactant. The opposite process is called quenching when a photoexited state is deactivated by a chemical reagent.

Most photochemical transformations occur through a series of simple steps known as primary photochemical processes. One common example of these processes is the excited state proton transfer (ESPT).

## Organic Photochemistry

Examples of photochemical organic reactions are electrocyclic reactions, photoisomerization and Norrish reactions.

Alkenes undergo many important reactions that proceed via a photon-induced ∂ to ∂* transition. The first electronic excited state of an alkene lack the ∂-bond, so that rotation about the C-C bond is rapid and the molecule engages in reactions not observed thermally. These reactions include cis-trans isomerization, cycloaddition to other (ground state) alkene to give cyclobutane derivatives. The cis-trans isomerization of a (poly)alkene is involved in retinal, a component of the machinery of vision. The dimerization of alkenes is relevant to the photodamage of DNA, where thymine dimers are observed upon illuminating DNA to UV radiation. Such dimers interfere with transcription. The beneficial effects of sunlight are associated with the photochemically induced retro-cyclisation (decyclisation) reaction of ergosterol to give vitamin D. In the DeMayo reaction, an alkene reacts with a 1,3-diketone reacts via its enol to yield a 1,5-diketone. Still another common photochemical reaction is Zimmerman's Di-pi-methane rearrangement.

In an industrial application, about 100,000 tonnes of benzyl chloride are prepared annually by the gas-phase photochemical reaction of toluene with chlorine. The light is absorbed by chlorine molecule, the low energy of this transition being indicted by the yellowish colour of the gas. The photon induces homolysis of the Cl-Cl bond, and the resulting chlorine radical converts toluene to the benzyl radical:

$$Cl_2 + h\mu \rightarrow 2\ Cl\cdot$$

$$C_6H_5CH_3 + Cl\cdot \rightarrow C_6H_5CH_2\cdot + HCl$$

$$C_6H_5CH_2\cdot + Cl\cdot \rightarrow C_6H_5CH_2Cl$$

Mercaptans can be produced by photochemical addition of hydrogen sulfide ($H_2S$) to alpha olefins.

### *Inorganic and Organometallic Photochemistry*

Coordination complexes and organometallic compounds are also photoreactive. These reactions can entail cis-trans isomerization. More commonly photoreactions result in dissociation of ligands, since the photon excites an electron on the metal to an orbital that is antibonding with respect to the ligands. Thus, metal carbonyls that resist thermal substitution undergo decarbonylation upon irradiation with UV light. UV-irradiation of a THF solution of molybdenum hexacarbonyl gives the THF complex, which is synthetically useful:

$$Mo(CO)_6 + THF \rightarrow Mo(CO)_5(THF) + CO$$

In a related reaction, photolysis of iron pentacarbonyl affords diiron nonacarbonyl:

$$2\ Fe(CO)_5 \rightarrow Fe_2(CO)_9 + CO$$

### *Carbon Nanotubes*

Certain species of Carbon Nanotubes also undergo photochemical reactions when exposed to intense pulsed light from a laser, arc lamp, or flashbulb. In the presence of air, Carbon Nanotubes will undergo combustion at high temperature (700-1500C). Under inert atmospheres, the nanotubes will break open and reorganise into larger "nanohorn" structures.

### *History*

Although bleaching has long been practiced, the first photochemical reaction was described by Trommsdorf in 1834. He observed that crystals of the compound α-santonin when exposed to sunlight turned yellow and burst.

In a 2007 study the reaction was described as a succession of three steps taking place within a single crystal.

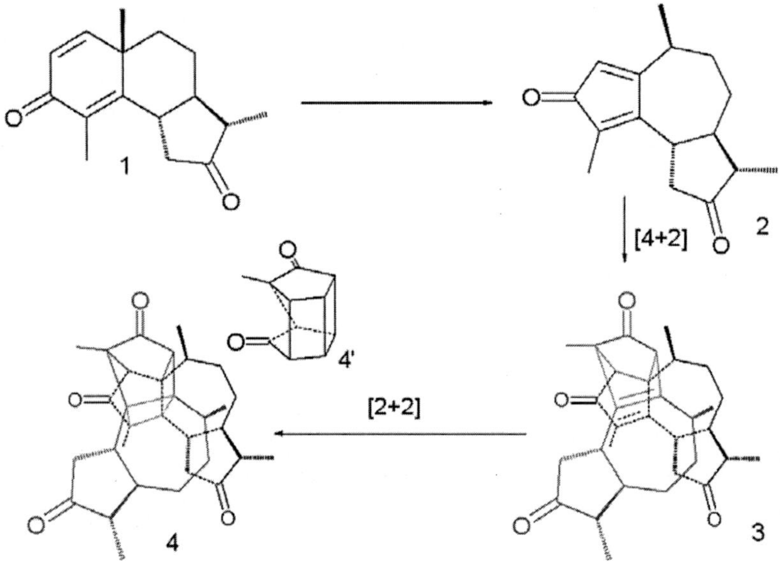


**Figure 2:** The first step is a rearrangement reaction to a cyclopentadienone intermediate 2, the second one a dimerization in a Diels-Alder reaction (3) and the third one a intramolecular [2+2]cycloaddition (4). The bursting effect is attributed to a large change in crystal volume on dimerization.

## Dendrimer

Dendrimers are repetitively branched molecules. The name comes from the Greek word (pronounced dendron), which translates to "tree". Synonymous terms for dendrimer include arborols and cascade molecules.

However, dendrimer is currently the internationally accepted term. A dendrimer is typically symmetric around the core, and often adopts a spherical three-dimensional morphology. The word dendron is also encountered frequently. A dendron usually contains a single chemically addressable group called the focal point.

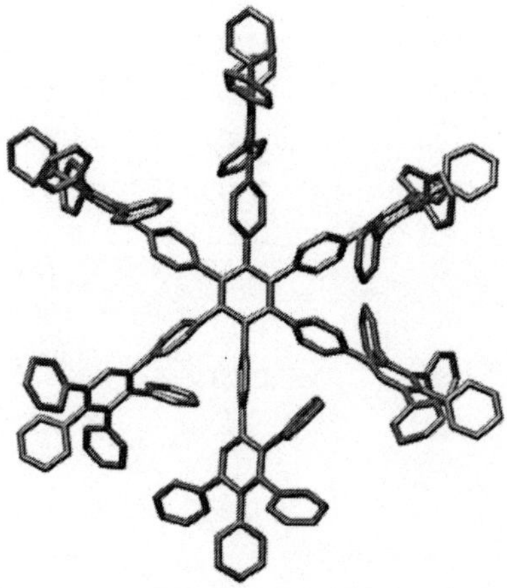

**Figure 3:** Crystal structure of a first-generation polyphenylene dendrimer reported by Müllen et al.

The first dendrimers were made by divergent synthesis approaches by Fritz in 1978, R.G. Denkewalter at Allied Corporation in 1981, Donald Tomalia at Dow Chemical in 1983 and in 1985, and by George Newkome in 1985. In 1990 a convergent synthetic approach was introduced by Jean Fréchet. Dendrimer popularity then greatly increased, resulting in more than 5,000 scientific papers and patents by the year 2005.

### Properties

Dendritic molecules are characterised by structural perfection. Dendrimers and dendrons are monodisperse and usually highly symmetric, spherical compounds. The field of dendritic molecules can be roughly divided into low-molecular weight and high-molecular weight species. The first category includes dendrimers and dendrons, and the latter includes dendronized polymers, hyperbranched polymers, and the polymer brush.

The properties of dendrimers are dominated by the functional groups on the molecular surface, however, there are examples of dendrimers with internal functionality. Dendritic encapsulation of functional molecules allows for the isolation of the active site, a structure that mimics that of active sites in biomaterials. Also, it is possible to make dendrimers water soluble, unlike mostpolymers, by functionalising their outer shell with charged species or other hydrophilic groups. Other controllable properties of dendrimers include toxicity, crystallinity, tecto-dendrimer formation, and chirality. Dendrimers are also classified by generation, which refers to the number of repeated branching cycles that are performed during its synthesis. For example if a dendrimer is made by convergent synthesis, and the branching reactions are performed onto the core molecule three times, the resulting dendrimer is considered a third generation dendrimer. Each successive generation results in a dendrimer roughly twice the molecular weight of the previous generation. Higher generation dendrimers also have more exposed functional groups on the surface, which can later be used to customize the dendrimer for a given application.

### *Synthesis*

One of the very first dendrimers, the Newkome dendrimer, was synthesized in 1985. This macromolecule is also commonly known by the name arborol. The synthesis is started by nucleophilic substitution of 1-bromopentane by *triethyl sodiomethanetricarboxylate* indimethylformamide and benzene.

The ester groups were then reduced by lithium aluminium hydride to a triol in a deprotection step. Activation of the chain ends was achieved by converting the alcohol groups to tosylate groups with tosyl chloride and pyridine. The tosyl group then served as leaving groups in another reaction with the tricarboxylate, forming generation two. Further repetition of the two steps leads to higher generations of arborol.

Poly(amidoamine), or PAMAM, is perhaps the most well known dendrimer. The core of PAMAM is a diamine (commonly ethylenediamine), which is reacted with methyl acrylate, and then another ethylenediamine to make the generation-0 (G-0) PAMAM. Successive reactions create higher generations, which tend to have different properties. Lower generations can be thought of as flexible molecules with no appreciable inner regions, while medium sized (G-3 or G-4) do have internal space that is essentially separated from the outer shell of the dendrimer. Very large (G-7 and greater) dendrimers can be thought of more like solid particles with very dense surfaces due to the structure of their outer shell.

Dendrimers can be considered to have three major portions: a core, an inner shell, and an outer shell. Ideally, a dendrimer can be synthesized to have different functionality in each of these portions to control properties such as solubility, thermal stability, and attachment of compounds for particular applications. Synthetic processes can also precisely control the size and number of branches on the dendrimer.

There are two defined methods of dendrimer synthesis, divergent synthesis and convergent synthesis. However, because the actual reactions consist of many steps needed to protect the active site, it is difficult to synthesize dendrimers using either method.

This makes dendrimers hard to make and very expensive to purchase. At this time, there are only a few companies that sell dendrimers; Polymer Factory Sweden AB commercialises biocompatible bis-MPA dendrimers and Dendritech is the only kilogram-scale producers of PAMAM dendrimers.

Dendritic Nanotechnologies Inc., from Mount Pleasant, Michigan, USA produces PAMAM dendrimers and other proprietary dendrimers.

### *Divergent Methods*

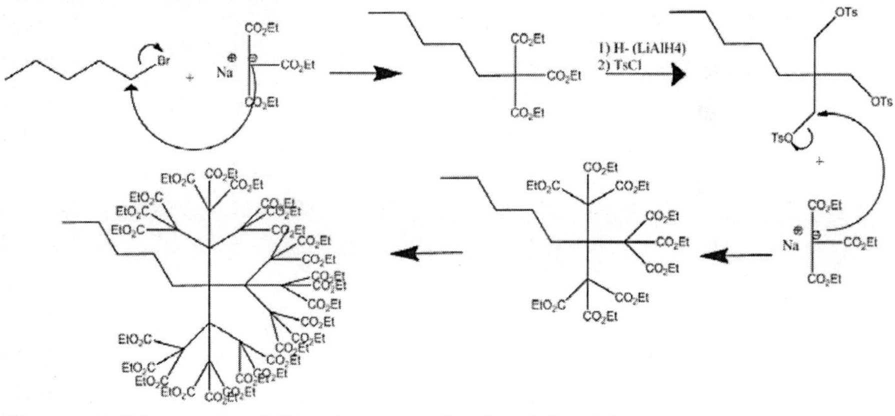

Figure 4: Schematic of divergent synthesis of dendrimers

The dendrimer is assembled from a multifunctional core, which is extended outward by a series of reactions, commonly a Michael reaction.

Each step of the reaction must be driven to full completion to prevent mistakes in the dendrimer, which can cause trailing generations (some branches are shorter than the others). Such impurities can impact the functionality and symmetry of the dendrimer, but are extremely difficult to purify out because the relative size difference between perfect and imperfect dendrimers is very small.

### Convergent Methods

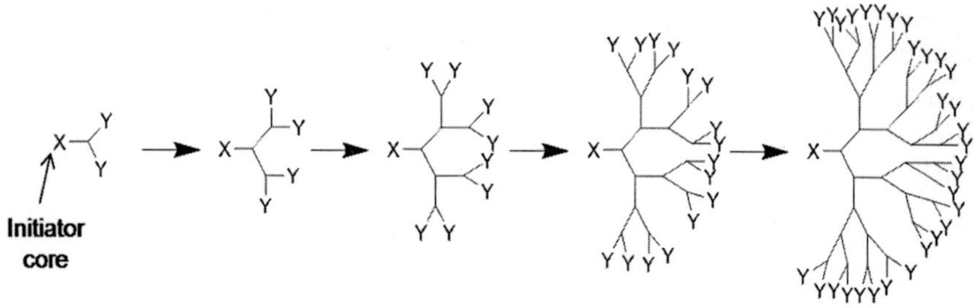


**Figure 5:** Schematic of convergent synthesis of dendrimers

Dendrimers are built from small molecules that end up at the surface of the sphere, and reactions proceed inward building inward and are eventually attached to a core. This method makes it much easier to remove impurities and shorter branches along the way, so that the final dendrimer is more monodisperse. However dendrimers made this way are not as large as those made by divergent methods because crowding due to steric effects along the core is limiting.

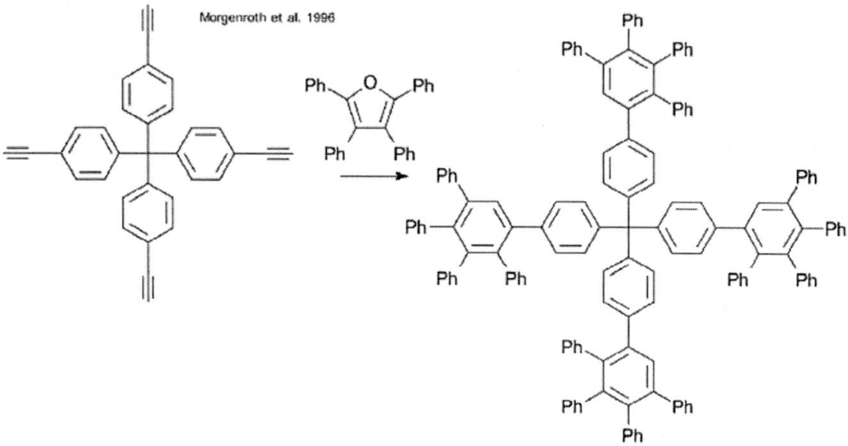


**Figure 6:** Dendrimer DA reaction Mullen 1996

There are ample avenues that can be opened by exploring this chemistry in dendrimer synthesis.

### Applications

Applications of dendrimers typically involve conjugating other chemical species to the dendrimer surface that can function as detecting agents (such as a dye molecule), affinity ligands, targeting components, radioligands, imaging agents, or pharmaceutically active compounds. Dendrimers have very strong potential for these applications because their structure can lead to multivalent systems. In other words, one dendrimer molecule has hundreds of possible sites to couple to an active species. Researchers aimed to utilise the hydrophobic environments of the dendritic media to conduct photochemical reactions that generate the products that are synthetically challenged. Carboxylic acid and phenol terminated water soluble dendrimers were synthesized to establish their utility in drug delivery as well as conducting chemical reactions in their interiors. This might allow researchers to attach both targeting molecules and drug molecules to the same dendrimer, which could reduce negative side effects of medications on healthy cells.

Dendrimers can also be used as a solubilizing agent. Since their introduction in the mid-1980s, this novel class of dendrimer architecture has been a prime candidate for hosts guest chemistry. Dendrimers with hydrophobic core and hydrophilic periphery have shown to exhibit micelle-like behaviour and have container properties in solution. The use of dendrimers as unimolecular micelles was proposed by Newkome in 1985. This analogy highlighted the utility of dendrimers as solubilizing agents. The majority of drugs available in pharmaceutical industry are hydrophobic in nature and this property in particular creates major formulation problems. This drawback of drugs can be ameliorated by dendrimeric scaffolding, which can be used to encapsulate as well as to solubilize the drugs because of the capability of such scaffolds to participate in extensive hydrogen bonding with water. Dendrimer labs throughout the planet are persistently trying to manipulate dendrimer's solubilizing trait, in their way to explore dendrimer as drug delivery and target specific carrier.

### *Drug Delivery*

Approaches for delivering unaltered natural products using polymeric carriers is of widespread interest, dendrimers have been explored for the encapsulation of hydrophobic compounds and for the delivery of anticancer drugs. The physical characteristics of dendrimers, including their monodispersity, water solubility, encapsulation ability, and large number of functionalisable peripheral groups, make these macromolecules appropriate candidates for evaluation as drug delivery vehicles. There are three methods for using dendrimers in drug delivery: first, the drug is covalently attached to the periphery of the dendrimer to form dendrimer prodrugs, second the drug is coordinated to the outer functional groups via ionic interactions, or third the dendrimer acts as a unimolecular micelle by encapsulating a pharmaceutical through the formation of a dendrimer-drug supramolecular assembly.

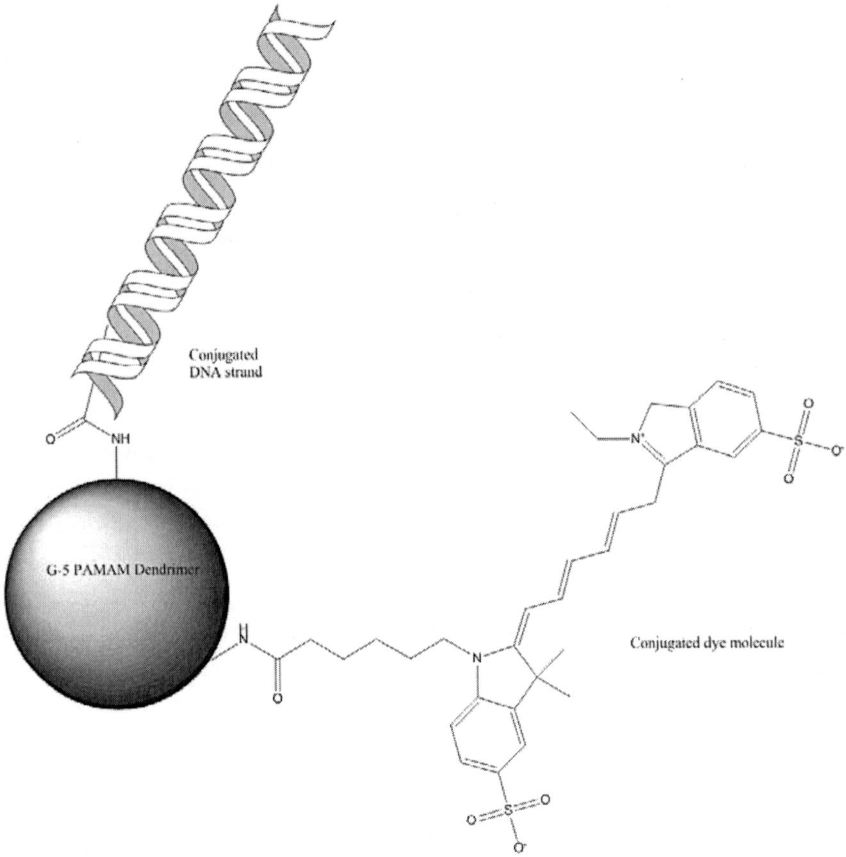


**Figure 7:** Schematic of a G-5 PAMAM dendrimer conjugated to both a dye molecule and a strand of DNA.

The use of dendrimers as drug carriers by encapsulating hydrophobic drugs is a potential method for delivering highly active pharmaceutical compounds that may not be in clinical use due to their limited water solubility and resulting suboptimal pharmacokinetics. Dendrimers have been widely explored for controlled delivery of antiretroviral bioactives The inherent antiretroviral activity of dendrimers enhances their efficacy as carriers for antiretroviral drugs The dendrimer enhances both the uptake and retention of compounds within cancer cells, a finding that was not anticipated at the onset of studies. The encapsulation increases with dendrimer generation and this method may be useful to entrap drugs with a relatively high therapeutic dose. Studies based on this dendritic polymer also open up new avenues of research into the further development of drug-dendrimer complexes specific for a cancer and/or targeted organ system. These encouraging results provide further impetus to design, synthesize, and evaluate dendritic polymers for use in basic drug delivery studies and eventually in the clinic.

### *Gene Delivery*

The ability to deliver pieces of DNA to the required parts of a cell includes many challenges. Current research is being performed to find ways to use dendrimers to traffic genes into cells without damaging or deactivating the DNA. To maintain the activity of DNA during dehydration, the dendrimer/DNA complexes were encapsulated in a water soluble polymer, and then deposited on or sandwiched in functional polymer films with a fast degradation rate to mediate gene transfection. Based on this method, PAMAM dendrimer/DNA complexes were used to encapsulate functional biodegradable polymer films for substratemediated gene delivery. Research has shown that the fast degrading functional polymer has great potential for localised transfection.

### *Sensors*

Scientists have also studied dendrimers for use in sensor technologies. Studied systems include proton or pH sensors using poly (propylene imine), cadmium-sulfide/polypropylenimine tetrahexacontaamine dendrimer composites to detect fluorescence signal quenching, and poly (propylenamine) first and second generation dendrimers for metal cation photodetection amongst others. Research in this field is vast and ongoing due to the potential for multiple detection and binding sites in dendritic structures.

## Nanoparticle

In nanotechnology, a particle is defined as a small object that behaves as a whole unit in terms of its transport and properties. Particles are further classified according to size : in terms of diametre, coarse particles cover a range between 10,000 and 2,500 nanometres. Fine particles are sized between 2,500 and 100 nanometres. Ultrafine particles, or nanoparticles are sized between 100 and 1 nanometres. The reason for this double name of the same object is that, during the 1970-80's, when the first thorough fundamental studies were running with "nanoparticles" in the USA (by Granqvist and Buhrman) and Japan, (within an ERATO Project) they were called "ultrafine particles" (UFP). However, during the 1990s before the National Nanotechnology Initiative was launched in the USA, the new name, "nanoparticle" had become fashionable. Nanoparticles may or may not exhibit size-related properties that differ significantly from those observed in fine particles or bulk materials. Although the size of most molecules would fit into the above outline, individual molecules are usually not referred to as nanoparticles.

Nanoclusters have at least one dimension between 1 and 10 nanometres and a narrow size distribution. Nanopowders are agglomerates of ultrafine particles, nanoparticles, or nanoclusters. Nanometre-sized single crystals, or single-domainultrafine particles, are often referred to as nanocrystals.

Nanoparticle research is currently an area of intense scientific interest due to a wide variety of potential applications in biomedical, optical and electronic fields.

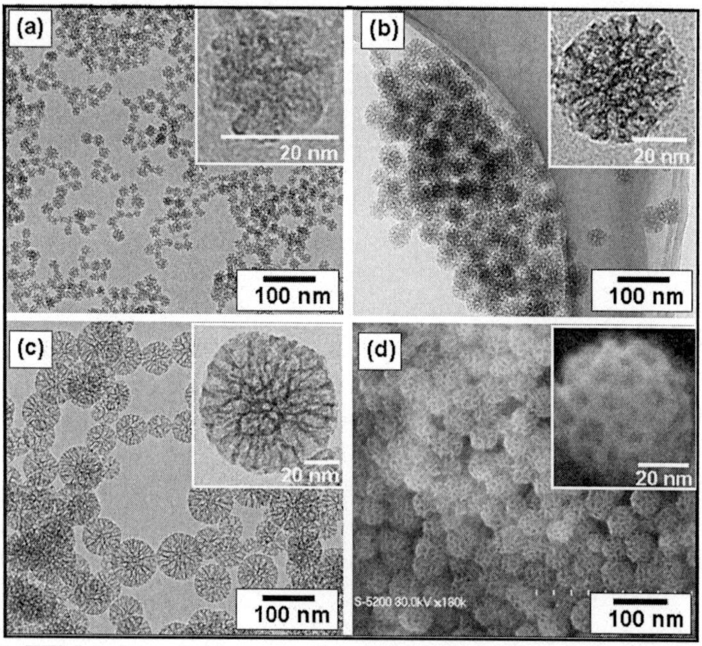


**Figure 8:** TEM (a, b, and c) images of prepared mesoporous silica nanoparticles with mean outer diametre: (a) 20nm, (b) 45nm, and (c) 80nm. SEM (d) image corresponding to (b). The insets are a high magnification of mesoporous silica particle.

The National Nanotechnology Initiative has led to generous public funding for nanoparticle research in the United States.

### Background

Although nanoparticles are generally considered a discovery of modern science, they actually have a very long history. Nanoparticles were used by artisans as far back as the 9th century in Mesopotamia for generating a glittering effect on the surface of pots.

Even these days, pottery from the Middle Ages and Renaissance often retain a distinct gold or copper coloured metallic glitter. This so called luster is caused by a metallic film that was applied to the transparent surface of a glazing. The luster can still be visible if the film has resisted atmospheric oxidation and other weathering.

The luster originated within the film itself, which contained silver and copper nanoparticles dispersed homogeneously in the glassy matrix of the ceramic glaze. These nanoparticles were created by the artisans by adding copper and silver salts and oxides together with vinegar, ochre and clay, on the surface of previously-glazed pottery. The object was then placed into a kiln and heated to about 600 °C in a reducing atmosphere. In the heat the glaze would soften, causing the copper and silver ions to migrate into the outer layers of the glaze. There the reducing atmosphere reduced the ions back to metals, which then came together forming the nanoparticles that give the colour and optical effects.

Luster technique showed that ancient craftsmen had a rather sophisticated empirical knowledge of materials. The technique originated in the islamic world. As Muslims were not allowed to use gold in artistic representations, they had to find a way to create a similar effect without using real gold. The solution they found was using luster.

Michael Faraday provided the first description, in scientific terms, of the optical properties of nanometre-scale metals in his classic 1857 paper. In a subsequent paper, the author (Turner) points out that: "It is well known that when thin leaves of gold or silver are mounted upon glass and heated to a temperature which is well below a red heat (~500 °C), a remarkable change of properties takes place, whereby the continuity of the metallic film is destroyed. The result is that white light is now freely transmitted, reflection is correspondingly diminished, while the electrical resistivity is enormously increased."

### *Uniformity*

The chemical processing and synthesis of high performance technological components for the private, industrial and military sectors requires the use of high purity ceramics, polymers, glass-ceramics and material composites. In condensed bodies formed from fine powders, the irregular particle sizes and shapes in a typical powder often lead to non-uniform packing morphologies that result in packing density variations in the powder compact.

Uncontrolled agglomeration of powders due to attractive van der Waals forces can also give rise to in microstructural inhomogeneities. Differential stresses that develop as a result of non-uniform drying shrinkage are directly related to the rate at which the solvent can be removed, and thus highly dependent upon the distribution of porosity. Such stresses have been associated with a plastic-to-brittle transition in consolidated bodies, and can yield to crack propagation in the unfired body if not relieved.

In addition, any fluctuations in packing density in the compact as it is prepared for the kiln are often amplified during the sintering process, yielding inhomogeneous densification. Some pores and other structural defects associated with density variations have been shown to play a detrimental role in the sintering process by growing and thus limiting end-point densities. Differential stresses arising from inhomogeneous densification have also been shown to result in the propagation of internal cracks, thus becoming the strength-controlling flaws.

Inert gas evaporation and inert gas deposition are free many of these defects due to the distillation (c.f. purification) nature of the process and having enough time to form single crystal particles, however even their non-aggregated deposits have lognormal size distribution, which is typical with nanoparticles. The reason why modern gas evaporation techniques can produce a relatively narrow size distribution is that aggregation can be avoided. However, even in this case, random residence times in the growth zone, due to the combination of drift and diffusion, result in a size distribution appearing lognormal.

It would therefore appear desirable to process a material in such a way that it is physically uniform with regard to the distribution of components and porosity, rather than using particle size distributions which will maximise the green density. The containment of a uniformly dispersed assembly of strongly interacting particles in suspension requires total control over interparticle forces. Monodisperse nanoparticles and colloids provide this potential.

Monodisperse powders of colloidal silica, for example, may therefore be stabilised sufficiently to ensure a high degree of order in the colloidal crystal or polycrystalline colloidal solid which results from aggregation. The degree of order appears to be limited by the time and space allowed for longer-range correlations to be established. Such defective polycrystalline colloidal structures would appear to be the basic elements of submicrometre colloidal materials science, and, therefore, provide the first step in developing a more rigorous understanding of the mechanisms involved in microstructural evolution in high performance materials and components.

### *Properties*

Nanoparticles are of great scientific interest as they are effectively a bridge between bulk materials and atomic or molecular structures. A bulk material should have constant physical properties regardless of its size, but at the nano-scale size-dependent properties are often observed. Thus, the properties of materials change as their size approaches the nanoscale and as the percentage of atoms at the surface of a material becomes significant. For bulk materials larger than one micrometre (or micron), the percentage of atoms at the surface is insignificant in relation to the number of atoms in the bulk of the material. *The interesting and sometimes unexpected properties of nanoparticles are therefore largely due to the large surface area of the material, which dominates the contributions made by the small bulk of the material.*

Nanoparticles often possess unexpected optical properties as they are small enough to confine their electrons and produce quantum effects. For example gold nanoparticles appear deep red to black in solution. Nanoparticles of usually yellow gold and gray silicon are red in colour. Gold nanoparticles melt at much lower temperatures (~300 °C for 2.5 nm size) than the gold slabs (1064 °C);. And absorption of solar radiation in photovoltaic cells is much higher in materials composed of nanoparticles than it is in thin films of continuous sheets of material. I.E. the smaller the particles, the greater the solar absorption.

Other size-dependent property changes include quantum confinement in semiconductor particles, surface plasmon resonance in some metal particles and superparamagnetism inmagnetic materials. Ironically, the changes in physical properties are not always desirable. Ferromagnetic materials smaller than 10 nm can switch their magnetisation direction using room temperature thermal energy, thus making them unsuitable for memory storage.

Suspensions of nanoparticles are possible since the interaction of the particle surface with the solvent is strong enough to overcome density differences, which otherwise usually result in a material either sinking or floating in a liquid.

The high surface area to volume ratio of nanoparticles provides a tremendous driving force for diffusion, especially at elevated temperatures. Sintering can take place at lower temperatures, over shorter time scales than for larger particles. This theoretically does not affect the density of the final product, though flow difficulties and the tendency of nanoparticles to agglomerate complicates matters. Moreover, nanoparticles have been found to impart some extra properties to various day to day products. For example the presence of titanium dioxide nanoparticles imparts what we call the self-cleaning effect, and the size being nano-range, the particles can not be observed. Zinc oxide particles have been found to have superior UV blocking properties compared to its bulk substitute. This is one of the reasons why it is often used in the preparation of sunscreen lotions, and is completely photostable.

Clay nanoparticles when incorporated into polymer matrices increase reinforcement, leading to stronger plastics, verifiable by a higher glass transition temperature and other mechanical property tests. These nanoparticles are hard, and impart their properties to the polymer (plastic). Nanoparticles have also been attached to textile fibers in order to create smart and functional clothing.

Metal, dielectric, and semiconductor nanoparticles have been formed, as well as hybrid structures (e.g., core-shell nanoparticles). Nanoparticles made of semiconducting material may also be labelled quantum dots if they are small enough (typically sub 10 nm) that quantization of electronic energy levels occurs. Such nanoscale particles are used in biomedical applications as drug carriers or imaging agents.

Semi-solid and soft nanoparticles have been manufactured. A prototype nanoparticle of semi-solid nature is the liposome. Various types of liposome nanoparticles are currently used clinically as delivery systems for anticancer drugs and vaccines.

Nanoparticles with one half hydrophilic and the other half hydrophobic are termed Janus particles and are particularly effective for stabilising emulsions. They can self-assemble at water/oil interfaces and act as solid surfactants.

### *Synthesis*

There are several methods for creating nanoparticles, including both attrition and pyrolysis. In attrition, macro or micro scale particles are ground in a ball mill, a planetary ball mill, or other size reducing mechanism. The resulting particles areair classified to recover nanoparticles. In pyrolysis, a vaporous precursor (liquid or gas) is forced through an orifice at high pressure and burned. The resulting solid (a version of soot) is air classified to recover oxide particles from by-product gases. Pyrolysis often results in aggregates and agglomerates rather than single primary particles.

A thermal plasma can also deliver the energy necessary to cause evaporation of small micrometre size particles. The thermal plasma temperatures are in the order of 10,000 K, so that solid powder easily evaporates. Nanoparticles are formed upon cooling while exiting the plasma region. The main types of the thermal plasma torches used to produce nanoparticles are dc plasma jet, dc arc plasma and radio frequency (RF) induction plasmas. In the arc plasma reactors, the energy necessary for evaporation and reaction is provided by an electric arc which is formed between the anode and the cathode. For example, silica sand can be vapourised with an arc plasma at atmospheric pressure. The resulting mixture of plasma gas and silica vapour can be rapidly cooled by quenching with oxygen, thus ensuring the quality of the fumed silica produced.

In RF induction plasma torches, energy coupling to the plasma is accomplished through the electromagnetic field generated by the induction coil. The plasma gas does not come in contact with electrodes, thus eliminating possible sources of contamination and allowing the operation of such plasma torches with a wide range of gases including inert, reducing, oxidising and other corrosive atmospheres. The working frequency is typically between 200 kHz and 40 MHz. Laboratory units run at power levels in the order of 30–50 kW while the large scale industrial units have been tested at power levels up to 1 MW. As the residence time of the injected feed droplets in the plasma is very short it is important that the droplet sizes are small enough in order to obtain complete evaporation. The RF plasma method has been used to synthesize different nanoparticle materials, for example synthesis of various ceramic nanoparticles such as oxides, carbours/carbides and nitrides of Ti and Si.

Inert-gas condensation is frequently used to make nanoparticles from metals with low melting points. The metal is vapourised in a vacuum chamber and then supercooled with an inert gas stream. The supercooled metal vapor condenses into nanometre-sized particles, which can be entrained in the inert gas stream and deposited on a substrate or studied in situ.

### *Sol-gel*

The sol-gel process is a wet-chemical technique (also known as chemical solution deposition) widely used recently in the fields of materials science and ceramic engineering. Such methods are used primarily for the fabrication of materials (typically a metal oxide) starting from a chemical solution (*sol*, short for solution) which acts as the precursor for an integrated network (or *gel*) of either discrete particles or network polymers.

Typical precursors are metal alkoxides and metal chlorides, which undergo hydrolysis and polycondensation reactions to form either a network "elastic solid" or a colloidal suspension (or dispersion) – a system composed of discrete (often amorphous) submicrometre particles dispersed to various degrees in a host fluid. Formation of a metal oxide involves connecting the metal centres with oxo (M-O-M) or hydroxo (M-OH-M) bridges, therefore generating metal-oxo or metal-hydroxo polymers in solution. Thus, the sol evolves towards the formation of a gel-like diphasic system containing both a liquid phase and solid phase whose morphologies range from discrete particles to continuous polymer networks.

In the case of the colloid, the volume fraction of particles (or particle density) may be so low that a significant amount of fluid may need to be removed initially for the gel-like properties to be recognised. This can be accomplished in any number of ways. The most simple method is to allow time for sedimentation to occur, and then pour off the remaining liquid. Centrifugation can also be used to accelerate the process of phase separation.

Removal of the remaining liquid (solvent) phase requires a drying process, which is typically accompanied by a significant amount of shrinkage and densification. The rate at which the solvent can be removed is ultimately determined by the distribution of porosity in the gel. The ultimate microstructure of the final component will clearly be strongly influenced by changes implemented during this phase of processing. Afterwards, a thermal treatment, or firing process, is often necessary in order to favour further polycondensation and enhance mechanical properties and structural stability via final sintering, densification and grain growth. One of the distinct advantages of using this methodology as opposed to the more traditional processing techniques is that densification is often achieved at a much lower temperature.

The precursor sol can be either deposited on a substrate to form a film (e.g. by dip-coating or spin-coating), cast into a suitable container with the desired shape (e.g. to obtain a monolithic ceramics, glasses, fibers, membranes, aerogels), or used to synthesize powders (e.g. microspheres, nanospheres). The sol-gel approach is a cheap and low-temperature technique that allows for the fine control of the product's chemical composition. Even small quantities of dopants, such as organic dyes and rare earth metals, can be introduced in the sol and end up uniformly dispersed in the final product. It can be used in ceramics processing and manufacturing as an investment casting material, or as a means of producing verythin films of metal oxides for various purposes. Sol-gel derived materials have diverse applications in optics, electronics, energy, space, biosensors, medicine (e.g. controlled drug release) and separation (e.g. chromatography) technology.

### *Colloids*

The term colloid is used primarily to describe a broad range of solid–liquid (and/or liquid–liquid) mixtures, all of which contain distinct solid (and/or liquid) particles which are dispersed to various degrees in a liquid medium. The term is specific to the size of the individual particles, which are larger than atomic dimensions but small enough to exhibit Brownian motion. If the particles are large enough, then their dynamic behaviour in any given period of time in suspension would be governed by forces of gravity and sedimentation. But if they are small enough to be colloids, then their irregular motion in suspension can be attributed to the collective bombardment of a myriad of thermally agitated molecules in the liquid suspending medium, as described originally by Albert Einstein in his dissertation. Einstein proved the existence of water molecules by concluding that this erratic particle behaviour could adequately be described using the theory of Brownian motion, with sedimentation being a possible long-term result. This critical size range (or particle diametre) typically ranges from nanometres (10"9 m) to micrometres (10"6 m).

### *Morphology*

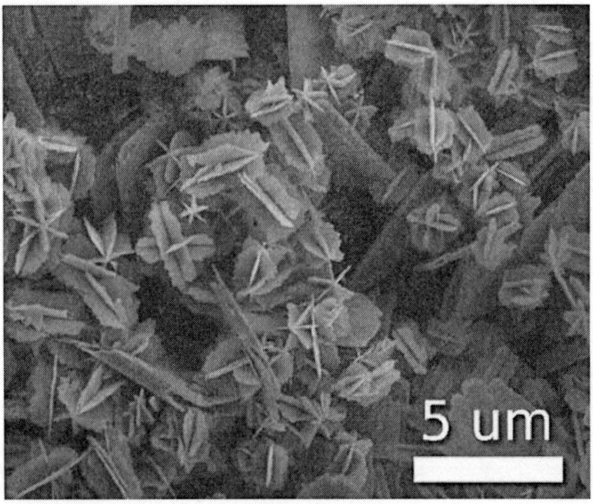


**Figure 9:** Nanostars of vanadium(IV) oxide

Scientists have taken to naming their particles after the real world shapes that they might represent. Nanospheres, nanoreefs, nanoboxes and more have appeared in the literature. These morphologies sometimes arise spontaneously as an effect of a templating or directing agent present in the synthesis such as miscellar emulsions or anodized alumina pores, or from the innate crystallographic growth patterns of the materials themselves. Some of these morphologies may serve a purpose, such as long carbon nanotubes being used to bridge an electrical junction, or just a scientific curiosity like the stars shown at right.

Amorphous particles usually adopt a spherical shape (due to their microstructural isotropy) – whereas the shape of anisotropic microcrystalline whiskers corresponds to their particular crystal habit. At the small end of the size range, nanoparticles are often referred to as clusters. Spheres, rods, fibers, and cups are just a few of the shapes that have been grown. The study of fine particles is called micromeritics.

### Characterisation

Nanoparticle characterisation is necessary to establish understanding and control of nanoparticle synthesis and applications. Characterisation is done by using a variety of different techniques, mainly drawn from materials science. Common techniques are electron microscopy (TEM, SEM), atomic force microscopy (AFM), dynamic light scattering (DLS), x-ray photoelectron spectroscopy (XPS), powder X-ray diffraction (XRD), Fourier transform infrared spectroscopy (FTIR), matrix-assisted laser desorption/ionization time-of-flight mass spectrometry (MALDI-TOF), ultraviolet-visible spectroscopy, dual polarisation interferometry and nuclear magnetic resonance (NMR).

Whilst the theory has been known for over a century, the technology for Nanoparticle tracking analysis (NTA) allows direct tracking of the Brownian motion and this method therefore allows the sizing of individual nanoparticles in solution.

### Functionalisation

The surface coating of nanoparticles is crucial to determining their properties. In particular, the surface coating can regulate stability, solubility and targeting. A coating that is multivalent or polymeric confers high stability.

### Surface Coating for Biological Applications

For biological applications, the surface coating should be polar to give high aqueous solubility and prevent nanoparticle aggregation. In serum or on the cell surface, highly charged coatings promote non-specific binding, while polyethylene glycol linked to terminal hydroxyl or methoxy groups repel non-specific interactions. Nanoparticles can be linked to biological molecules which can act as address tags, to direct the nanoparticles to specific sites within the body, specific organelles within the cell, or to follow specifically the movement of individual protein or RNA molecules in living cells. Common address tags are monoclonal antibodies, aptamers, streptavidin or peptides. These targeting agents should ideally be covalently linked to the nanoparticle and should be present in a controlled number per nanoparticle. Multivalent nanoparticles, bearing multiple targeting groups, can cluster receptors, which can activate cellular signalling pathways, and give stronger anchoring. Monovalent nanoparticles, bearing a single binding site, avoid clustering and so are preferable for tracking the behaviour of individual proteins.

### Safety

Nanoparticles present possible dangers, both medically and environmentally. Most of these are due to the high surface to volume ratio, which can make the particles very reactive or catalytic. They are also able to pass through cell membranes in organisms, and their interactions with biological systems are relatively unknown. A recent study looking at the effects of ZnO nanoparticles on human immune cells has found varying levels of susceptibility to cytotoxicity. Smaller nanoparticles evinced increased cytotoxicity. Lymphocytes (especially naive T cells) were found to be more resistant to nanoparticle cytotoxicity than monocytes, likely due to the capacity of the latter to produce higher levels of reactive oxygen species in response to internalised nanoparticles. Previously activated memory T cells were more susceptible than naive T cells, implying a relationship between cell-cycle and nanoparticle susceptibility. In addition, nanoparticle concentrations below those causing appreciable cell death nonetheless induced the production of proinflammatory cytokines, such as IFN-$\gamma$ and TNF. Despite these laboratory findings, free nanoparticles in the environment may rapidly agglomerate and thus leave the nano-regime. Nature itself presents many nanoparticles to which organisms on earth may have evolved immunity (such as salt particulates from ocean aerosols, terpenes from plants, or dust from volcanic eruptions). A more complete analysis is provided in the article on nanotechnology.

Animal studies have shown that some nanoparticles can penetrate cells and tissues, move through the body and brain and cause biochemical damage. They also have shown to cause a risk factor in men for testicular cancer. But whether cosmetics and sunscreens containing nanomaterials pose health risks remains largely unknown at this stage. Diesel nanoparticles have been found to damage the cardiovascular system in a mouse model.

Concern has also been raised over the health effects of respirable nanoparticles from certain combustion processes.

### Laser Applications

The use of nanoparticle distributions in laser dye-doped poly (methyl methacrylate) (PMMA) laser gain media was demonstrated in 2003 and it has been shown to improve conversion efficiencies and to decrease laser beam divergence. Researchers attribute the reduction in beam divergence to improved dn/dT characteristics of the organic-inorganic dye-doped nanocomposite. The optimum composition reported by these researchers is 30% w/w of $SiO_2$ (~ 12 nm) in dye-doped PMMA.

## Fullerene

A fullerene is any molecule composed entirely of carbon, in the form of a hollow sphere, ellipsoid, or tube. Spherical fullerenes are also called buckyballs, and they resemble the balls used in association football. Cylindrical ones are called carbon nanotubes or buckytubes. Fullerenes are similar in structure to graphite, which is composed of stacked graphene sheets of linked hexagonal rings; but they may also contain pentagonal (or sometimes heptagonal) rings.

The first fullerene to be discovered, and the family's namesake, buckminsterfullerene ($C_{60}$), was prepared in 1985 by Richard Smalley, Robert Curl, James Heath, Sean ,O'Brien, and Harold Kroto at Rice University. The name was an homage to Buckminster Fuller, whose geodesic domes it resembles. The structure was also identified some five years earlier by Sumio Iijima, from an electron microscope image, where it formed the core of a "bucky onion." Fullerenes have since been found to occur in nature. More recently, fullerenes have been detected in outer space. According to astronomer Letizia Stanghellini, "It's possible that buckyballs from outer space provided seeds for life on Earth."

The discovery of fullerenes greatly expanded the number of known carbon allotropes, which until recently were limited to graphite, diamond, and amorphous carbon such as soot and charcoal. Buckyballs and buckytubes have been the subject of intense research, both for their unique chemistry and for their technological applications, especially in materials science, electronics, and nanotechnology.

### *History*

The icosahedral $C_{60}H_{60}$ cage was mentioned in 1965 as a possible topological structure. The existence of $C_{60}$ was predicted by Eiji Osawa of Toyohashi University of Technology in 1970. He noticed that the structure of a corannulene molecule was a subset of football shape, and he hypothesised that a full ball shape could also exist. His idea was reported in Japanese scientific journals, but did not reach Europe or the Americas.

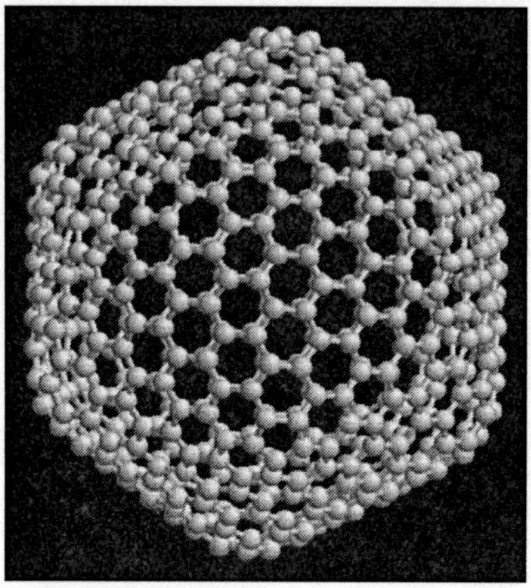

**Figure 10:** The icosahedral fullerene $C_{540}$, another member of the family of fullerenes.

Also in 1970, R. W. Henson (then of the Atomic Energy Research Establishment) proposed the structure and made a model of $C_{60}$. Unfortunately, the evidence for this new form of carbon was very weak and was not accepted, even by his colleagues. The results were never published but were acknowledged in *Carbon* in 1999.

Independently from Henson in 1973 the group of scientists from USSR directed by Prof. Bochvar made the quantum-chemical analysis of stability of $C_{60}$ and calculated electronic structure of the molecule. As in the last cases the theoretical prediction wasn't accepted by scientific community. The paper was published in 1973 in *Proceedings of the USSR Academy of Sciences* (in Russian).

In mass spectrometry, discrete peaks appeared corresponding to molecules with the exact mass of sixty or seventy or more carbon atoms. In 1985, Harold Kroto (then of the University of Sussex), James R. Heath, Sean O'Brien, Robert Curl and Richard Smalley, from Rice University, discovered $C_{60}$, and shortly thereafter came to discover the fullerenes. Kroto, Curl, and Smalley were awarded the 1996 Nobel Prize in Chemistry for their roles in the discovery of this class of molecules. $C_{60}$ and other fullerenes were later noticed occurring outside the laboratory (e.g., in normal candle soot). By 1991, it was relatively easy to produce gram-sized samples of fullerene powder using the techniques of Donald Huffman, Wolfgang KrΔtschmer and Konstantinos Fostiropoulos. Fullerene purification remains a challenge to chemists and to a large extent determines fullerene prices. So-called endohedral fullerenes have ions or small molecules incorporated inside the cage atoms. Fullerene is an unusual reactant in manyorganic reactions such as the Bingel reaction discovered in 1993. Carbon nanotubes were recognised in 1991.

Minute quantities of the fullerenes, in the form of $C_{60}$, $C_{70}$, $C_{76}$, and $C_{84}$ molecules, are produced in nature, hidden in soot and formed by lightning discharges in the atmosphere. In 1992, fullerenes were found in a family of minerals known as Shungites in Karelia, Russia.

In 2010, fullerenes ($C_{60}$) have been discovered in a cloud of cosmic dust surrounding a distant star 6500 light years away. Using NASA's Spitzer infrared telescope the scientists spotted the molecules' unmistakable infrared signature. Sir Harry Kroto, who shared the 1996 Nobel Prize in Chemistry for the discovery of buckyballs commented: "This most exciting breakthrough provides convincing evidence that the buckyball has, as I long suspected, existed since time immemorial in the dark recesses of our galaxy."

### *Naming*

Buckminsterfullerene ($C_{60}$) was named after Richard Buckminster Fuller, a noted architectural modeler who popularised the geodesic dome. Since buckminsterfullerenes have a shape similar to that sort of dome, the name was thought appropriate. As the discovery of the fullerene family came *after* buckminsterfullerene, the shortened name 'fullerene' is used to refer to the family of fullerenes. The suffix "ene" indicates that each C atom is covalently bonded to three others (instead of the maximum of four), a situation that classically would correspond to the existence of bonds involving two pairs of electrons ("double bonds").

### *Types of Fullerene*

Since the discovery of fullerenes in 1985, structural variations on fullerenes have evolved well beyond the individual clusters themselves. Examples include:

- buckyball clusters: smallest member is $C_{20}$ (unsaturated version of dodecahedrane) and the most common is $C_{60}$;
- nanotubes: hollow tubes of very small dimensions, having single or multiple walls; potential applications in electronics industry;
- megatubes: larger in diametre than nanotubes and prepared with walls of different thickness; potentially used for the transport of a variety of molecules of different sizes;

- polymers: chain, two-dimensional and three-dimensional polymers are formed under high-pressure high-temperature conditions
- nano"onions": spherical particles based on multiple carbon layers surrounding a buckyball core; proposed for lubricants;
- linked "ball-and-chain" dimers: two buckyballs linked by a carbon chain;
- fullerene rings.

### Buckyballs

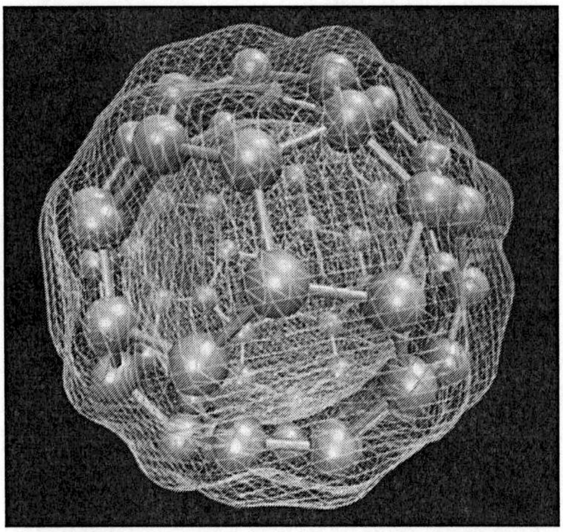

**Figure 11:** $C_{60}$ with isosurface of ground state electron density as calculated with DFT

**Figure 12:** Many association footballshave the same shape as buckminsterfullerene, $C_{60}$

### Buckminsterfullerene

Buckminsterfullerene is the smallest fullerene molecule in which no two pentagons share an edge (which can be destabilising, as in pentalene). It is also the most common in terms of natural occurrence, as it can often be found in soot.

The structure of $C_{60}$ is a truncated (T = 3) icosahedron, which resembles an association football ball of the type made of twenty hexagons and twelve pentagons, with a carbon atom at the vertices of each polygon and a bond along each polygon edge.

The van der Waals diametre of a $C_{60}$ molecule is about 1.1 nanometres (nm). The nucleus to nucleus diametre of a $C_{60}$ molecule is about 0.71 nm.

The $C_{60}$ molecule has two bond lengths. The 6:6 ring bonds (between two hexagons) can be considered "double bonds" and are shorter than the 6:5 bonds (between a hexagon and a pentagon). Its average bond length is 1.4 angstroms.

Silicon buckyballs have been created around metal ions.

### Boron Buckyball

A type of buckyball which uses boron atoms, instead of the usual carbon, was predicted and described in 2007. The $B_{80}$ structure, with each atom forming 5 or 6 bonds, is predicted to be more stable than the $C_{60}$ buckyball. One reason for this given by the researchers is that the B-80 is actually more like the original geodesic dome structure popularised by Buckminster Fuller, which uses triangles rather than hexagons. However, this work has been subject to much criticism by quantum chemists as it was concluded that the predicted $I_h$ symmetric structure was vibrationally unstable and the resulting cage undergoes a spontaneous symmetry break, yielding a puckered cage with rare $T_h$ symmetry (symmetry of a volleyball). The number of six-member rings in this molecule is 20 and number of five-member rings is 12. There is an additional atom in the centre of each six-member ring, bonded to each atom surrounding it.

### Other Buckyballs

Another fairly common fullerene is $C_{70}$, but fullerenes with 72, 76, 84 and even up to 100 carbon atoms are commonly obtained.

In mathematical terms, the structure of a fullerene is a trivalent convex polyhedron with pentagonal and hexagonal faces.

In graph theory, the term fullerene refers to any 3-regular, planar graph with all faces of size 5 or 6 (including the external face).

It follows from Euler's polyhedron formula, *V* ” *E* + *F* = 2, (where *V*, *E*, *F* are the numbers of vertices, edges, and faces), that there are exactly 12 pentagons in a fullerene and *V*/2 ” 10 hexagons.

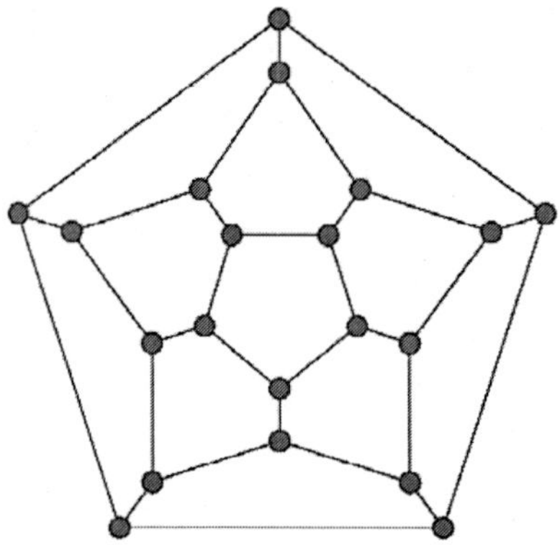

20-fullerene
(dodecahedral graph)

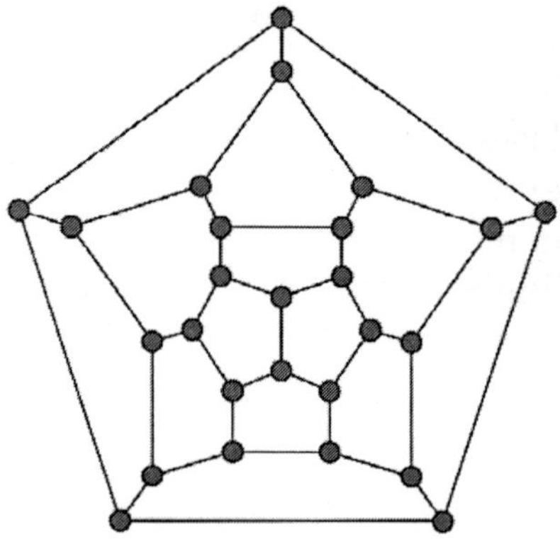

26-fullerene graph

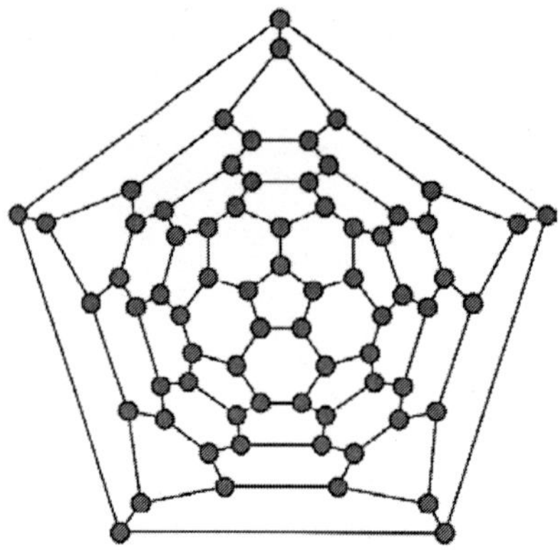

60-fullerene

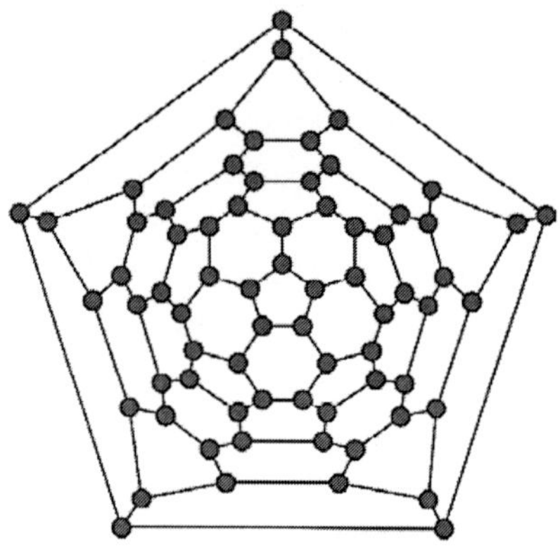

70-fullerene graph

**Figure 13:** Truncated icosahedral graph

The smallest fullerene is the dodecahedral $C_{20}$. There are no fullerenes with 22 vertices. The number of fullerenes $C_{2n}$ grows with increasing $n$ = 12, 13, 14, ..., roughly in proportion to $n^9$ (sequence A007894 in OEIS). For instance, there are 1812 non-isomorphic fullerenes $C_{60}$. Only one form of $C_{60}$, the buckminsterfullerene alias truncated icosahedron, has no pair of adjacent pentagons (the smallest such fullerene). To further illustrate the growth, there are 214,127,713 non-isomorphic fullerenes $C_{200}$, 15,655,672 of which have no adjacent pentagons.

### Trimetasphere

carbon nanomaterials were discovered by researchers at Virginia Tech and licensed exclusively to Luna Innovations. This class of novel molecules comprises 80 carbon atoms ($C_{80}$) forming a sphere which encloses a complex of three metal atoms and one nitrogen atom. These fullerenes encapsulate metals which puts them in the subset referred to as metallofullerenes. Trimetaspheres have the potential for use in diagnostics (as safe imaging agents), therapeutics and in organic solar cells.

### Carbon Nanotubes

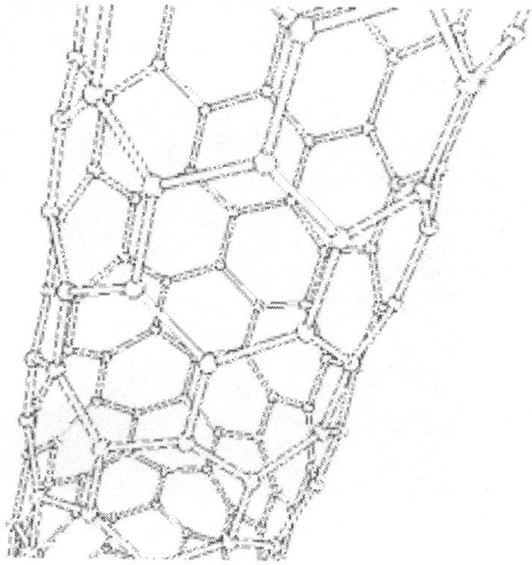

**Figure 14:** This rotating model of a carbon nanotube shows its 3D structure.

Nanotubes are cylindrical fullerenes. These tubes of carbon are usually only a few nanometres wide, but they can range from less than a micrometre to several millimetres in length. They often have closed ends, but can be open-ended as well. There are also cases in which the tube reduces in diametre before closing off. Their unique molecular structure results in extraordinary macroscopic properties, including high tensile strength, high electrical conductivity, high ductility, high heat conductivity, and relative chemical inactivity (as it is cylindrical and "planar" — that is, it has no "exposed" atoms that can be easily displaced). One proposed use of carbon nanotubes is in paper batteries, developed in 2007 by researchers at Rensselaer Polytechnic Institute.

Another highly speculative proposed use in the field of space technologies is to produce high-tensile carbon cables required by a space elevator.

### Carbon Nanobuds

Nanobuds have been obtained by adding buckminsterfullerenes to carbon nanotubes.

### Fullerite

Fullerites are the solid-state manifestation of fullerenes and related compounds and materials.

"Ultrahard fullerite" is a coined term frequently used to describe material produced by high-pressure high-temperature (HPHT) processing of fullerite. Such treatment converts fullerite into a nanocrystalline form of diamond which has been reported to exhibit remarkable mechanical properties.

**Figure 15:** The $C_{60}$ fullerene in crystalline form

### Properties

For the past decade, the chemical and physical properties of fullerenes have been a hot topic in the field of research and development, and are likely to continue to be for a long time. *Popular Science* has published articles about the possible uses of fullerenes in armour. In April 2003, fullerenes were under study for potential medicinal use: binding specific antibiotics to the structure to target resistant bacteria and even target certain cancer cells such as melanoma. The October 2005 issue of *Chemistry & Biology* contains an article describing the use of fullerenes as light-activated antimicrobial agents.

In the field of nanotechnology, heat resistance and superconductivity are some of the more heavily studied properties.

A common method used to produce fullerenes is to send a large current between two nearby graphite electrodes in an inert atmosphere. The resulting carbon plasma arc between the electrodes cools into sooty residue from which many fullerenes can be isolated.

There are many calculations that have been done using ab-initio quantum methods applied to fullerenes. By DFT and TD-DFT methods one can obtain IR, Raman and UV spectra. Results of such calculations can be compared with experimental results.

### *Aromaticity*

Researchers have been able to increase the reactivity of fullerenes by attaching active groups to their surfaces. Buckminsterfullerene does not exhibit "superaromaticity": that is, the electrons in the hexagonal rings do not delocalise over the whole molecule.

A spherical fullerene of $n$ carbon atoms has $n$ pi-bonding electrons, free to delocalise. These should try to delocalise over the whole molecule. The quantum mechanics of such an arrangement should be like one shell only of the well-known quantum mechanical structure of a single atom, with a stable filled shell for $n$ = 2, 8, 18, 32, 50, 72, 98, 128, etc.; i.e. twice a perfect square number; but this series does not include 60. This $2(N + 1)^2$ rule (with $N$ integer) for spherical aromaticity is the three-dimensional analogue of Hückel's rule. The 10+ cation would satisfy this rule, and should be aromatic. This has been shown to be the case using quantum chemical modelling, which showed the existence of strong diamagnetic sphere currents in the cation.

As a result, $C_{60}$ in water tends to pick up two more electrons and become an anion. The $nC_{60}$ described below may be the result of $C_{60}$ trying to form a loose metallic bond.

### *Chemistry*

Fullerenes are stable, but not totally unreactive. The $sp^2$-hybridised carbon atoms, which are at their energy minimum in planar graphite, must be bent to form the closed sphere or tube, which produces angle strain. The characteristic reaction of fullerenes is electrophilic addition at 6,6-double bonds, which reduces angle strain by changing $sp^2$-hybridised carbons into $sp^3$-hybridised ones. The change in hybridised orbitals causes the bond angles to decrease from about 120° in the $sp^2$ orbitals to about 109.5° in the $sp^3$ orbitals. This decrease in bond angles allows for the bonds to bend less when closing the sphere or tube, and thus, the molecule becomes more stable.

Other atoms can be trapped inside fullerenes to form inclusion compounds known as endohedral fullerenes. An unusual example is the egg shaped fullerene $Tb_3N@C_{84}$, which violates the isolated pentagon rule. Recent evidence for a meteor impact at the end of the Permian period was found by analysing noble gases so preserved. Metallofullerene-based inoculates using the rhonditic steel process are beginning production as one of the first commercially-viable uses of buckyballs.

### *Solubility*

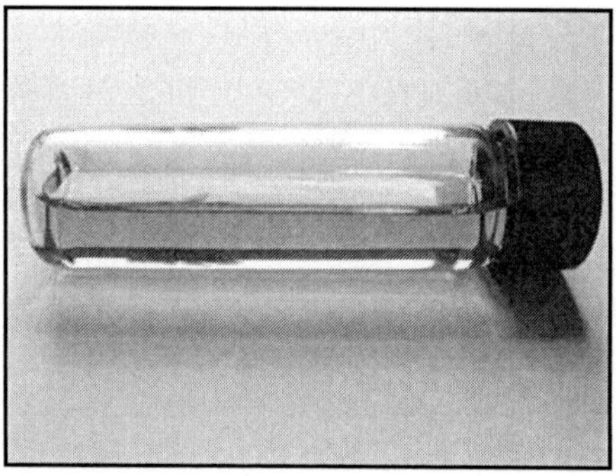

**Figure 16:** $C_{60}$ in solution

Fullerenes are sparingly soluble in many solvents. Common solvents for the fullerenes include aromatics, such as toluene, and others like carbon disulfide. Solutions of pure buckminsterfullerene have a deep purple colour. Solutions of $C_{70}$ are a reddish brown. The higher fullerenes $C_{76}$ to $C_{84}$ have a variety of colours. $C_{76}$ has two optical forms, while other higher fullerenes have several structural isomers. Fullerenes are the only known allotrope of carbon that can be dissolved in common solvents at room temperature.

| *Solvent* | $C_{60}$ | $C_{70}$ |
|---|---|---|
| 1-chloronaphthalene | 51 mg/mL | * |
| 1-methylnaphthalene | 33 mg/mL | * |
| 1,2-dichlorobenzene | 24 mg/mL | 36.2 mg/mL |
| 1,2,4-trimethylbenzene | 18 mg/mL | * |
| tetrahydronaphthalene | 16 mg/mL | * |
| carbon disulfide | 8 mg/mL | 9.875 mg/mL |
| 1,2,3-tribromopropane | 8 mg/mL | * |
| xylene | 5 mg/mL | 3.985 mg/mL(p-xylene) |
| bromoform | 5 mg/mL | * |

| cumene | 4 mg/mL | * |
|---|---|---|
| toluene | 3 mg/mL | 1.406 mg/mL |
| benzene | 1.5 mg/mL | 1.3 mg/mL |
| carbon tetrachloride | 0.447 mg/mL | 0.121 mg/mL |
| chloroform | 0.25 mg/mL | * |
| n-hexane | 0.046 mg/mL | 0.013 mg/mL |
| cyclohexane | 0.035 mg/mL | 0.08 mg/mL |
| tetrahydrofuran | 0.006 mg/mL | * |
| acetonitrile | 0.004 mg/mL | * |
| methanol | 0.000 04 mg/mL | * |
| water | $1.3\times10^{-11}$ mg/mL | * |
| pentane | 0.004 mg/mL | 0.002 mg/mL |
| heptane | * | 0.047 mg/mL |
| octane | 0.025 mg/mL | 0.042 mg/mL |
| isooctane | 0.026 mg/mL | * |
| decane | 0.070 mg/mL | 0.053 mg/mL |
| dodecane | 0.091 mg/mL | 0.098 mg/mL |
| tetradecane | 0.126 mg/mL | * |
| acetone | * | 0.0019 mg/mL |
| isopropanol | * | 0.0021 mg/mL |
| dioxane | 0.0041 mg/mL | * |
| mesitylene | 0.997 mg/mL | 1.472 mg/mL |
| dichloromethane | 0.254 mg/mL | 0.080 mg/mL |

* : Solubility not measured

Some fullerene structures are not soluble because they have a small band gap between the ground and excited states. These include the small fullerenes $C_{28}$, $C_{36}$ and $C_{50}$. The $C_{72}$ structure is also in this class, but the endohedral version with a trappedlanthanide-group atom is soluble due to the interaction of the metal atom and the electronic states of the fullerene. Researchers had originally been puzzled by $C_{72}$ being absent in fullerene plasma-generated soot extract, but found in endohedral samples. Small band gap fullerenes are highly reactive and bind to other fullerenes or to soot particles.

Solvents that are able to dissolve buckminsterfullerene ($C_{60}$ and $C_{70}$) are listed at left in order from highest solubility. The solubility value given is the approximate saturated concentration.

Solubility of $C_{60}$ in some solvents shows unusual behaviour due to existence of solvate phases (analogues of crystallohydrates). For example, solubility of $C_{60}$ in benzenesolution shows maximum at about 313 K. Crystallisation from benzene solution at temperatures below maximum results in formation of triclinic solid solvate with four benzene molecules $C_{60}\cdot 4C_6H_6$ which is rather unstable in air. Out of solution, this structure decomposes into usual fcc $C_{60}$ in few minutes' time. At temperatures above solubility maximum the solvate is not stable even when immersed in saturated solution and melts with formation of fcc $C_{60}$. Crystallisation at temperatures above the solubility maximum results in formation of pure fcc $C_{60}$. Millimetre-sized crystals of $C_{60}$ and $C_{70}$ can be grown from solution both for solvates and for pure fullerenes.

### *Hydrated Fullerene (HyFn)*

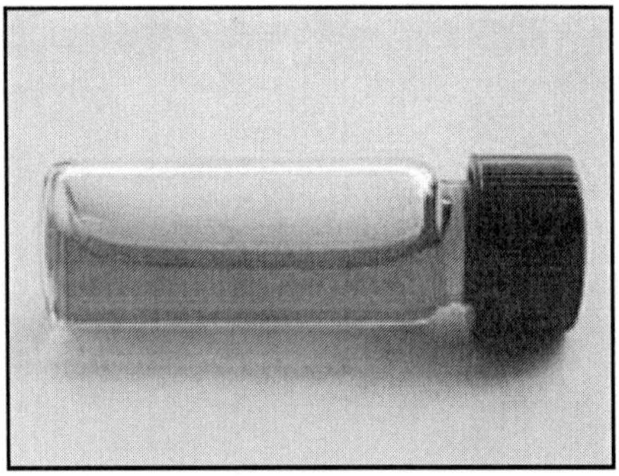

**Figure 17:** $C_{60}$HyFn water solution with a $C_{60}$ concentration of 0.22 g/L.

Hydrated fullerene $C_{60}$HyFn is a stable, highly hydrophilic, supramolecular complex consisting of $\tilde{N}_{60}$ fullerene molecule enclosed into the first hydrated shell that contains 24 water molecules: $C_{60}@(H_2O)_{24}$. This hydrated shell is formed as a result of donor-acceptor interaction between lone-electron pairs of oxygen, water molecules and electron-acceptor centres on the fullerene surface. Meanwhile, the water molecules which are oriented close to the fullerene surface are interconnected by a three-dimensional network of hydrogen bonds. The size of $C_{60}$HyFn is 1.6–1.8 nm. The maximal concentration of $\tilde{N}_{60}$ in the form of $C_{60}$HyFn achieved by 2010 is 4 mg/mL.

### *Quantum Mechanics*

In 1999, researchers from the University of Vienna demonstrated that wave-particle duality applied to molecules such as fullerene. One of the co-authors of this research, Julian Voss-Andreae, has since created several sculptures symbolizing wave-particle duality in fullerenes.

Science writer Marcus Chown stated on the CBC radio show *Quirks and Quarks* in May 2006 that scientists are trying to make buckyballs exhibit the quantum behaviour of existing in two places at once (quantum superposition).

## *Safety and Toxicity*

Moussa *et al.* (1996-7) studied the *in vivo* toxicity of $C_{60}$ after intra-peritoneal administration of large doses. No evidence of toxicity was found and the mice tolerated a dose of 5 000 mg/kg of body weight (BW). Mori *et al.* (2006) could not find toxicity in rodents for $C_{60}$ and $C_{70}$ mixtures after oral administration of a dose of 2 000 mg/kg BW and did not observe evidence of genotoxic or mutagenic potential *in vitro*. Other studies could not establish the toxicity of fullerenes: on the contrary, the work of Gharbi*et al.* (2005) suggested that aqueous $C_{60}$ suspensions failing to produce acute or subacute toxicity in rodents could also protect their livers in a dose-dependent manner against free-radical damage.

A comprehensive and recent review on fullerene toxicity is given by Kolosnjaj *et al.* (2007a,b, c). These authors review the works on fullerene toxicity beginning in the early 1990s to present, and conclude that very little evidence gathered since the discovery of fullerenes indicate that $C_{60}$ is toxic.

With reference to nanotubes, a recent study by Poland *et al.* (2008) on carbon nanotubes introduced into the abdominal cavity of mice led the authors to suggest comparisons to "asbestos-like pathogenicity". It should be noted that this was not an inhalation study, though there have been several performed in the past, therefore it is premature to conclude that nanotubes should be considered to have a toxicological profile similar to asbestos. Conversely, and perhaps illustrative of how the various classes of molecules which fall under the general term fullerene cover a wide range of properties, Sayes *et al.* found that *in vivo* inhalation of $C_{60}(OH)_{24}$ and nano-$C_{60}$ in rats gave no effect, whereas in comparison quartz particles produced an inflammatory response under the same conditions. As stated above, nanotubes are quite different in chemical and physical properties to $C_{60}$, i.e., molecular weight, shape, size, physical properties (such as solubility) all are very different, so from a toxicological standpoint, different results for $C_{60}$ and nanotubes are not suggestive of any discrepancy in the findings.

When considering toxicological data, care must be taken to distinguish as necessary between what are normally referred to as fullerenes: ($C_{60}$, $C_{70}$, ...); fullerene derivatives: $C_{60}$ or other fullerenes with covalently bonded chemical groups; fullerene complexes (e.g., water-solubilized with surfactants, such as $C_{60}$-PVP; host-guest complexes, such as with cyclodextrin), where the fullerene is physically bound to another molecule; $C_{60}$ nanoparticles, which are extended solid-phase aggregates of $C_{60}$ crystallites; and nanotubes, which are generally much larger (in terms of molecular weight and size) molecules, and are different in shape to the spheroidal fullerenes $C_{60}$ and $C_{70}$, as well as having different chemical and physical properties.

The above different molecules span the range from insoluble materials in either hydrophilic or lipophilic media, to hydrophilic, lipophilic, or even amphiphilic molecules, and with other varying physical and chemical properties. Therefore any broad generalisation extrapolating for example results from $C_{60}$ to nanotubes or vice versa is not possible, though technically all are fullerenes, as the term is defined as a close-caged all-carbon molecule. Any extrapolation of results from one molecule to other molecules must take into account considerations based on a quantitative structural analysis relationship study (QSARS), which mostly depends on how close the molecules under consideration are in physical and chemical properties.

### Superconductivity

After the synthesis of macroscopic amounts of fullerenes, their physical properties could be investigated. Haddon *et al.* found that intercalation of alkali-metal atoms in solid $C_{60}$ leads to metallic behaviour. In 1991, it was revealed that potassium-doped $C_{60}$ becomes superconducting at 18 K. This was the highest transition temperature for a molecular superconductor. Since then, superconductivity has been reported in fullerene doped with various other alkali metals. It has been shown that the superconducting transition temperature in alkaline-metal-doped fullerene increases with the unit-cell volume V. As caesium forms the largest alkali ion, caesium-doped fullerene is an important material in this family. Recently, superconductivity at 38 K has been reported in bulk $Cs_3C_{60}$, but only under applied pressure. The highest superconducting transition temperature of 33 K at ambient pressure is reported for $Cs_2RbC_{60}$.

The increase of transition temperature with the unit-cell volume had been believed to be evidence for the BCS mechanism of $C_{60}$ solid superconductivity, because inter $C_{60}$ separation can be related to an increase in the density of states on the Fermi level, $N(\varepsilon_F)$. Therefore, there have been many efforts to increase the interfullerene separation, in particular, intercalating neutral molecules into the $A_3C_{60}$ lattice to increase the interfullerene spacing while the valence of $C_{60}$ is kept unchanged. However, this ammoniation technique has revealed a new aspect of fullerene intercalation compounds: the Mott-Hubbard transition and the correlation between the orientation/orbital order of $C_{60}$ molecules and the magnetic structure.

The $C_{60}$ molecules compose a solid of weakly bound molecules. The fullerites are therefore molecular solids, in which the molecular properties still survive. The discrete levels of a free $C_{60}$ molecule are only weakly broadened in the solid, which leads to a set of essentially nonoverlapping bands with a narrow width of about 0.5 eV. For an undoped $C_{60}$ solid, the 5-fold $h_u$ band is the HOMO level, and the 3-fold $t_{1u}$ band is the empty LUMO level, and this system is a band insulator. But when the $C_{60}$ solid is doped with metal atoms, the metal atoms give electrons to the $t_{1u}$ band or the upper 3-fold $t_{1g}$ band. This partial electron occupation of the band leads to sometimes metallic behaviour. However, $A_4C_{60}$ is an insulator, although the $t_{1u}$ band is only partially filled and it should be a metal according to band theory. This unpredicted behaviour may be explained by the Jahn-Teller effect, where spontaneous deformations of high-symmetry molecules induce the splitting of degenerate levels to gain the electronic energy. The Jahn-Teller type electron-phonon interaction is strong enough in $C_{60}$ solids to destroy the band picture for particular valence states.

A narrow band or strongly correlated electronic system and degenerated ground states are important points to understand in explaining superconductivity in fullerene solids. When the inter-electron repulsion U is greater than the bandwidth, an insulating localised electron ground state is produced in the simple Mott-Hubbard model. This explains the absence of superconductivity at ambient pressure in caesium-doped $C_{60}$ solids. Electron-correlation-driven localisation of the $t_{1u}$ electrons exceeds the critical value, leading to the Mott insulator. The application of high pressure decreases the interfullerene spacing, therefore caesium-doped $C_{60}$ solids turn to metallic and superconducting. A fully developed theory of $C_{60}$ solids superconductivity is still lacking, but it has been widely accepted that strong electronic correlations and the Jahn-Teller electron-phonon coupling produce local electron-pairings that show a high transition temperature close to the insulator-metal transition.

### Chirality

Some fullerenes (e.g. $C_{76}$, $C_{78}$, $C_{80}$, and $C_{84}$) are inherently chiral because they are $D_2$-symmetric, and have been successfully resolved. Research efforts are ongoing to develop specific sensors for their enantiomers.

### Popular Culture

Examples of fullerenes in popular culture are numerous. Fullerenes appeared in fiction well before scientists took serious interest in them. In *New Scientist* there used to be a weekly column called "Daedalus" written by David Jones, which contained humorous descriptions of unlikely technologies. In 1966 Jones suggested that it may be possible to create giant hollow carbon molecules by distorting a plane hexagonal net by the addition of impurity atoms.

On 4 September 2010, Google used an interactively rotatable fullerene $C_{60}$ as the second 'o' in their logo to celebrate the 25th anniversary of the discovery of the fullerenes.

## Cell Membrane

The cell membrane or plasma membrane is a biological membrane that separates the interior of all cells from the outside environment. The cell membrane is selectively permeable to ions and organic molecules and controls the movement of substances in and out of cells. It basically protects the cell from outside forces. It consists of the lipid bilayer with embedded proteins. Cell membranes are involved in a variety of cellular processes such as cell adhesion, ion conductivity and cell signalling and serve as the attachment surface for several extracellular structures, including the cell wall, glycocalyx, and intracellular cytoskeleton.

### Function

The cell membrane surrounds the cytoplasm of a cell and, in animal cells, physically separates the intracellular components from the extracellular environment. Fungi, bacteria and plants also have the cell wall which provides a mechanical support for the cell and precludes the passage of larger molecules. The cell membrane also plays a role in anchoring the cytoskeleton to provide shape to the cell, and in attaching to the extracellular matrix and other cells to help group cells together to form tissues.

The membrane is differentially permeable and able to regulate what enters and exits the cell, thus facilitating the transport of materials needed for survival. The movement of substances across the membrane can be either *passive*, occurring without the input of cellular energy, or active, requiring the cell to expend energy in transporting it. The membrane also maintains the cell potential. The cell membrane thus works as a selective filter that allows only certain things to come inside or go outside the cell. To do so, the membrane employs a number of transport mechanisms:

1. Diffusion : Some substances (small molecules, ions) such as carbon dioxide (CO2), oxygen (O2), and water, can move across the plasma membrane by diffusion, which is a passive transport process.
2. Osmosis : Because the membrane acts as a barrier for certain molecules and ions, they can occur in different concentrations on the two sides of the membrane. Such a concentration difference across a semipermeable membrane can set up a osmotic flow for the solvent, in this case water. Water can thus be transported across the membrane by osmosis.
3. Mediated Transport : Nutrients such as sugars and materials of growth such as amino acid must enter the cell, and the waste of metabolism must leave. Such molecules are moved across the membrane by special proteins called transport proteins or permeases. Permeases form a small passageway through the membrane, enabling the solute molecule to cross the phospholipid bilayer. Permeases are usually quite specific, recognising and transporting only a limited group of chemical substances, often even only a single substance.
4. Endocytosis : Endocytosis is the process in which cells absorb molecules by engulfing them. The plasma membrane creates a small deformation inward, called an invagination, in which the substance to be transported is captured. The deformation then pinches off from the membrane on the inside of the cell,

creating a vesicle containing the captured substance. Endocytosis is a pathway for internalising solid particles (cell eating or phagocytosis), small molecules and ions (cell drinking or pinocytosis), and macromolecules. Endocytosis requires energy and is thus a form of active transport.

5. Exocytosis : Just as material can be brought into the cell by invagination and formation of a vesicle, the membrane of a vesicle can be fused with the plasma membrane, extruding its contents to the surrounding medium. This is the process of exocytosis. Exocytosis occurs in various cells to remove undigested residues of substances brought in by endocytosis, to secrete substances such as hormones and enzymes, and to transport a substance completely across a cellular barrier. In the process of exocytosis, the undigested waste-containing food vacuole or the secretory vesicle budded from Golgi apparatus, is first moved by cytoskeleton from the interior of the cell to the surface. The vesicle membrane comes in contact with the plasma membrane. The lipid molecules of the two bilayers rearrange themselves and the two membranes are, thus, fused. A passage is formed in the fused membrane and the vesicles discharges its contents outside the cell.

### Prokaryotes

Gram-negative bacteria have a plasma membrane and an outer membrane separated by a periplasmic space. Other prokaryotic species have only a plasma membrane. Prokaryotic cells are also surrounded by a cell wall composed of peptidoglycan (amino acid and sugar). Some eukaryotic cells also have cells walls, but none that are made of peptidoglycan.

### Structure

***Fluid Mosaic Model:*** According to the fluid mosaic model of S.J. Singer and G.L. Nicolson (1972), biological membranes can be considered as a two-dimensional liquid in which all lipid and protein molecules diffuse more or less easily. Although the lipid bilayers that form the basis of the membranes do indeed form two-dimensional liquids by themselves, the plasma membrane also contains a large quantity of proteins, which provide more structure. Examples of such structures are protein-protein complexes, pickets and fences formed by the actin-based cytoskeleton, and potentially lipid rafts.

### *Lipid Bilayer*

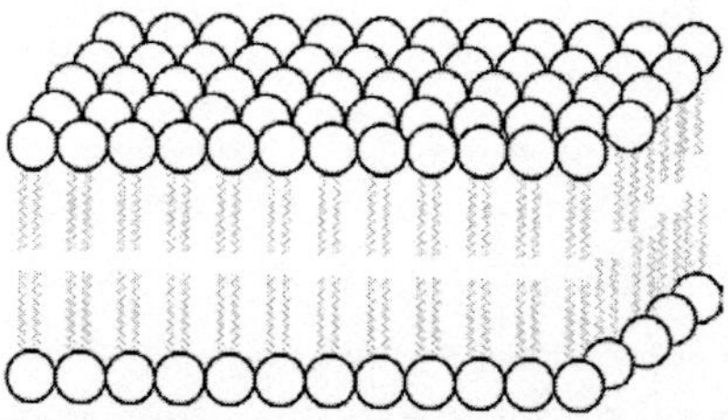

**Figure 18:** Diagram of the arrangement of amphipathic lipid molecules to form a lipid bilayer. The yellow polar head groups separate the grey hydrophobic tails from the aqueous cytosolic and extracellular environments.

Lipid bilayers go through a self assembly process in the formation of membranes. The cell membrane consists primarily of a thin layer of amphipathic phospholipids which spontaneously arrange so that the hydrophobic "tail" regions are shielded from the surrounding polar fluid, causing the more hydrophilic "head" regions to associate with the cytosolic and extracellular faces of the resulting bilayer. This forms a continuous, spherical lipid bilayer. Forces such as Van der Waal, electrostatic, hyrdogen bonds, and noncovalent interactions, are all forces that contribute to the formation of the lipid bilayer. Overall, hydrophobic interactions are the major driving force in the formation of lipid bilayers.

Lipid bilayers have very low permeability for ions and most polar molecules. The arrangement of hydrophilic heads and hydrophobic tails of the lipid bilayer prevent polar solutes (e.g. amino acids, nucleic acids, carbohydrates, proteins, and ions) from diffusing across the membrane, but generally allows for the passive diffusion of hydrophobic molecules. This affords the cell the ability to control the movement of these substances via transmembrane protein complexes such as pores and gates.

Flippases and scramblases concentrate phosphatidyl serine, which carries a negative charge, on the inner membrane. Along with NANA, this creates an extra barrier to charged moieties moving through the membrane.

Membranes serve diverse functions in eukaryotic and prokaryotic cells. One important role is to regulate the movement of materials into and out of cells. The phospholipid bilayer structure (fluid mosaic model) with specific membrane proteins accounts for the selective permeability of the membrane and passive and active transport mechanisms. In addition, membranes in prokaryotes and in the mitochondria and chloroplasts of eukaryotes facilitate the synthesis of ATP through chemiosmosis.

## *Membrane Polarity*

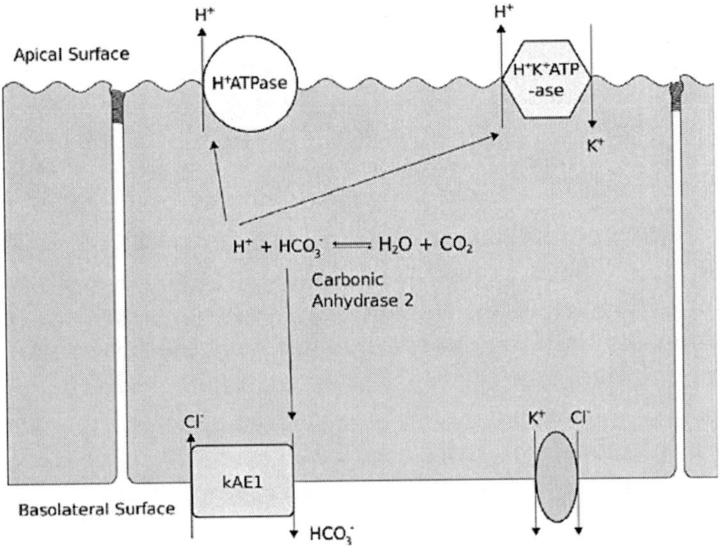


**Figure 19:** Alpha intercalated cell

The apical membrane of a polarised cell is the surface of the plasma membrane that faces the lumen. This is particularly evident in epithelial and endothelial cells, but also describes other polarised cells, such as neurons.

The basolateral membrane of a polarised cell is the surface of the plasma membrane that forms its basal and lateral surfaces. It faces towards the interstitium, and away from the lumen.

"Basolateral membrane" is a compound phrase referring to the terms *basal (base) membrane* and *lateral (side) membrane*, which, especially in epithelial cells, are identical in composition and activity. Proteins (such as ion channels and pumps) are free to move from the basal to the lateral surface of the cell or *vice versa* in accordance with the fluid mosaic model.

Tight junctions that join epithelial cells near their apical surface prevent the migration of proteins from the basolateral membrane to the apical membrane. The basal and lateral surfaces thus remain roughly equivalent to one another, yet distinct from the apical surface.

### *Integral Membrane Proteins*

The cell membrane contains many integral membrane proteins, which pepper the entire surface. These structures, which can be visualised by electron microscopy or fluorescence microscopy, can be found on the inside of the membrane, the outside, or may span the entire membrane. These may include integrins, cadherins, desmosomes, clathrin-coated pits, caveolaes, and different structures involved in cell adhesion. Integral proteins are the most abundant type of protein to span the lipid bilayer. They interact widely with hydrocarbon chains of membrane lipids and can be released by agents that compete for the same nonpolar interactions.

### *Peripheral Membrane Proteins*

Peripheral proteins are proteins that are bounded to the membrane by electrostatic interactions and hydrogen bonding with the hydrophilic phospholipid heads. Many of these proteins can be found bounded to the surfaces of integral proteins on either the cytoplasimic side of the cell or the extracellular side of the membrane. Some are anchored to the bilayer through covalent bond with a fatty acid.

### *Membrane Skeleton*

The cytoskeleton is found underlying the cell membrane in the cytoplasm and provides a scaffolding for membrane proteins to anchor to, as well as forming organelles that extend from the cell. Indeed, cytoskeletal elements interact extensively and intimately with the cell membrane. Anchoring proteins restricts them to a particular cell surface — for example, the *apical surface* of epithelial cells that line the vertebrate gut — and limits how far they may diffuse within the bilayer.

The cytoskeleton is able to form appendage-like organelles, such as cilia, which are microtubule-based extensions covered by the cell membrane, and filopodia, which are actin-based extensions. These extensions are ensheathed in membrane and project from the surface of the cell in order to sense the external environment and/or make contact with the substrate or other cells.

The apical surfaces of epithelial cells are dense with actin-based finger-like projections known as microvilli, which increase cell surface area and thereby increase the absorption rate of nutrients. Localised decoupling of the cytoskeleton and cell membrane results in formation of a bleb.

## *Composition*

Cell membranes contain a variety of biological molecules, notably lipids and proteins. Material is incorporated into the membrane, or deleted from it, by a variety of mechanisms:

- Fusion of intracellular vesicles with the membrane (exocytosis) not only excretes the contents of the vesicle but also incorporates the vesicle membrane's components into the cell membrane. The membrane may form blebs around extracellular material that pinch off to become vesicles (endocytosis).
- If a membrane is continuous with a tubular structure made of membrane material, then material from the tube can be drawn into the membrane continuously.
- Although the concentration of membrane components in the aqueous phase is low (stable membrane components have low solubility in water), there is an exchange of molecules between the lipid and aqueous phases.

## *Lipids*

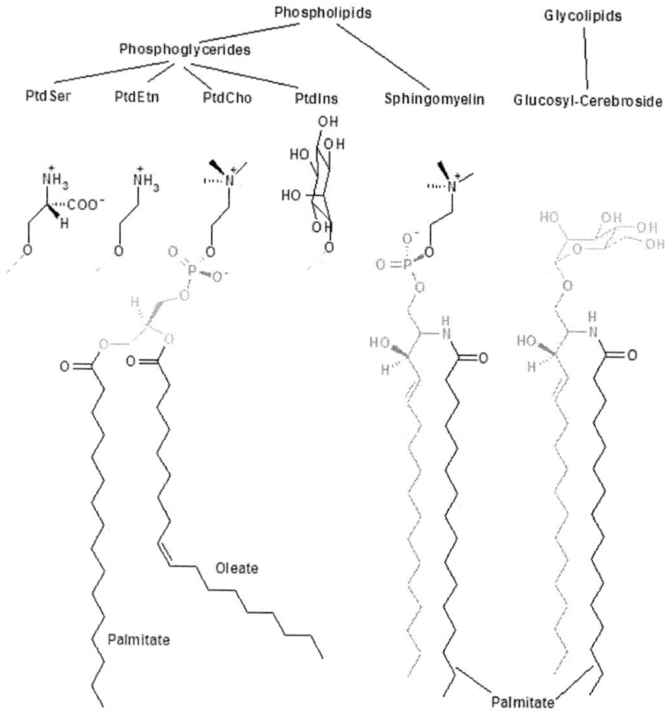


**Figure 20:** Examples of the major membrane phospholipids and glycolipids: phosphatidylcholine (PtdCho), phosphatidylethanolamine (PtdEtn), phosphatidylinositol (PtdIns), phosphatidylserine (PtdSer).

The cell membrane consists of three classes of amphipathic lipids: phospholipids, glycolipids, and cholesterols. The amount of each depends upon the type of cell, but in the majority of cases phospholipids are the most abundant. In RBC studies, 30% of the plasma membrane is lipid.

The fatty chains in phospholipids and glycolipids usually contain an even number of carbon atoms, typically between 16 and 20. The 16- and 18-carbon fatty acids are the most common. Fatty acids may be saturated or unsaturated, with the configuration of the double bonds nearly always *cis*. The length and the degree of unsaturation of fatty acid chains have a profound effect on membrane fluidity as unsaturated lipids create a kink, preventing the fatty acids from packing together as tightly, thus decreasing the melting temperature (increasing the fluidity) of the membrane. The ability of some organisms to regulate the fluidity of their cell membranes by altering lipid composition is called homeoviscous adaptation.

The entire membrane is held together via non-covalent interaction of hydrophobic tails, however the structure is quite fluid and not fixed rigidly in place. Under physiological conditions phospholipid molecules in the cell membrane are in the liquid crystalline state. It means the lipid molecules are free to diffuse and exhibit rapid lateral diffusion along the layer in which they are present. However, the exchange of phospholipid molecules between intracellular and extracellular leaflets of the bilayer is a very slow process. Lipid rafts and caveolae are examples of cholesterol-enriched microdomains in the cell membrane.

In animal cells cholesterol is normally found dispersed in varying degrees throughout cell membranes, in the irregular spaces between the hydrophobic tails of the membrane lipids, where it confers a stiffening and strengthening effect on the membrane.

### *Phospholipids Forming Lipid Vesicles*

Lipid vesicles or lisosomes are circular pockets that are enclosed by a lipid bilayer. These structures are used in laboratories to study the effects of chemicals in cells by delivering these chemicals directly to the cell, as well as getting more insight into cell membrane permeability. Lipid vesicles and liposomes are formed by first suspending a lipid in an aqueous solution then agitating the mixture through sonication, resulting in a vesicle. By measuring the rate of efflux from that of the inside of the vesicle to the ambient solution, allows researcher to better understand membrane permeability. Vesicles can be formed with molecules and ions inside the vesicle by forming the vesicle with the desired molecule or ion present in the solution. Proteins can also be embedded into the membrane through solubilizing the desired proteins in the presence of detergents and attaching them to the phospholipids in which the liposome is formed. These provide researchers with a tool to examine various membrane protein functions.

### Carbohydrates

Plasma membranes also contain carbohydrates, predominantly glycoproteins, but with some glycolipids (cerebrosides and gangliosides). For the most part, no glycosylation occurs on membranes within the cell; rather generally glycosylation occurs on the extracellular surface of the plasma membrane. The glycocalyx is an important feature in all cells, especially epithelia with microvilli. Recent data suggest the glycocalyx participates in cell adhesion, lymphocyte homing, and many others. The penultimate sugar is galactose and the terminal sugar is sialic acid, as the sugar backbone is modified in the golgi apparatus. Sialic acid carries a negative charge, providing an external barrier to charged particles.

### Proteins

Proteins within the membrane are key to the functioning of the overall membrane. These proteins mainly transport chemicals and information across the membrane. Every membrane has a varying degree of protein content. Proteins can be in the form of peripheral or integral.

The cell membrane, being exposed to the outside environment, is an important site of cell-cell communication. As such, a large variety of protein receptors and identification proteins, such as antigens, are present on the surface of the membrane. Functions of membrane proteins can also include cell-cell contact, surface recognition, cytoskeleton contact, signalling, enzymatic activity, or transporting substances across the membrane.

Most membrane proteins must be inserted in some way into the membrane. For this to occur, an N-terminus “signal sequence” of amino acids directs proteins to the endoplasmic reticulum, which inserts the proteins into a lipid bilayer. Once inserted, the proteins are then transported to their final destination in vesicles, where the vesicle fuses with the target membrane.

### *Variation*

The cell membrane has different lipid and protein compositions in distinct types of cells and may have therefore specific names for certain cell types:

- Sarcolemma in myocytes
- Oolemma in oocytes
- Historically, the plasma membrane was also referred to as the plasmalemma.

### *Permeability*

The permeability of a membrane is the ease with which molecules pass through it. These molecules are known as permeant molecules. Permeability depends mainly on the electric charge of the molecule and to a lesser extent the molar mass of the molecule. Due to the cell membrane's hydrophobic nature, electrically neutral and small molecules pass through the membrane easier than charged, large ones.

The inability of charged molecules to pass through the cell membrane results in pH parturition of substances throughout the fluid compartments of the body.

## Vesicle (Biology and Chemistry)

A vesicle is a bubble of liquid within another liquid, a supramolecular assembly made up of many different molecules. More technically, a vesicle is a small membrane-enclosed sack that can store or transport substances. Vesicles can form naturally because of the properties of lipid membranes, or they may be prepared. Artificially prepared vesicles are known as liposomes. Most vesicles have specialised functions depending on what materials they contain.

Because vesicles tend to look alike, it is very difficult to tell the difference between different types.

The vesicle is separated from the cytosol by at least one phospholipid bilayer. If there is only one phospholipid bilayer, they are called *unilamellar* vesicles; otherwise they are called *multilamellar*.

Vesicles store, transport, or digest cellular products and waste. The membrane enclosing the vesicle is similar to that of the plasma membrane, and vesicles can fuse with the plasma membrane to release their contents outside of the cell. Vesicles can also fuse with other organelles within the cell.

Because it is separated from the cytosol, the inside of the vesicle can be made to be different from the cytosolic environment. For this reason, vesicles are a basic tool used by the cell for organising cellular substances. Vesicles are involved in metabolism, transport, buoyancy control, and enzyme storage. They can also act as chemical reaction chambers.

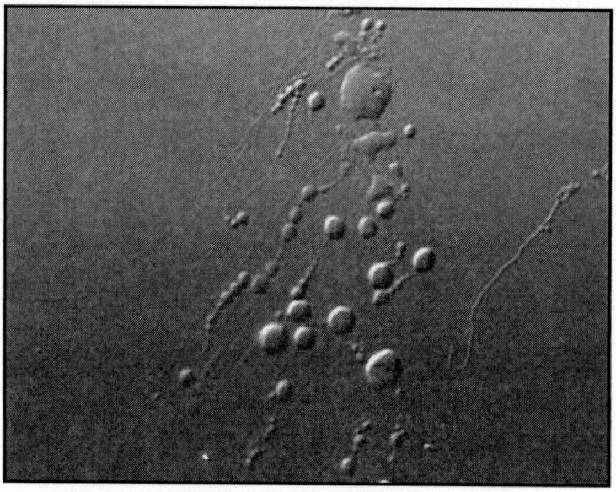

**Figure 21:** Sarfus image of lipid vesicles.

### *Types of Vesicles*

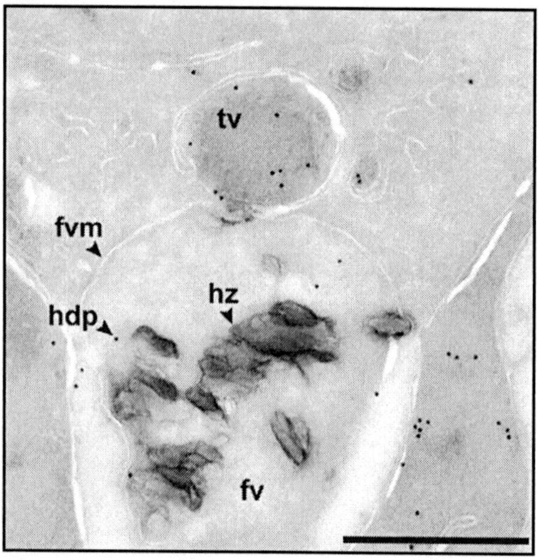


**Figure 22:** Electron micrograph of a cell containing a food vacuole (fv) and transport vacuole (tv) in a malaria parasite.

### *Vacuoles*

Vacuoles are vesicles which contain mostly water.

- Plant cells are known for having a *large central vacuole* in the centre of the cell that is used for osmotic control and nutrient storage.
- Contractile vacuoles are found in certain protists, especially those in Phylum *Ciliophora.* These vacuoles take water from the cytoplasm and excrete it from the cell to avoid bursting due to osmotic pressure.

### *Lysosomes*

- Lysosomes or "Suicide Cells" are involved in cellular digestion. Food can be taken from outside the cell into food vacuoles by a process called endocytosis. These food vacuoles fuse with lysosomes which break down the components so that they can be used in the cell. This form of cellular eating is called phagocytosis.
- Lysosomes are also used to destroy defective or damaged organelles in a process called endophagocytosis. They fuse with the membrane of the damaged organelle digesting it.

### *Transport Vesicles*

- Transport vesicles can move molecules between locations inside the cell, e.g., proteins from the rough endoplasmic reticulum to the Golgi apparatus.
- Membrane-bound and secreted proteins are made on ribosomes found in the rough endoplasmic reticulum. Most of these proteins mature in the Golgi apparatus before going to their final destination which may be to lysosomes, peroxisomes, or outside of the cell. These proteins travel within the cell inside of transport vesicles.

### *Secretory Vesicles*

Secretory vesicles contain materials that are to be excreted from the cell. Cells have many reasons to excrete materials. One reason is to dispose of wastes.

Another reason is tied to the function of the cell. Within a larger organism, some cells are specialised to produce certain chemicals. These chemicals are stored in secretory vesicles and released when needed. Some examples include the following.

### Types of Secretory Vesicles

- Synaptic vesicles are located at presynaptic terminals in neurons and store neurotransmitters. When a signal comes down an axon, the synaptic vesicles fuse with the cell membrane releasing the neurotransmitter so that it can be detected by receptor molecules on the next nerve cell.
- In animals endocrine tissues release hormones into the bloodstream. These hormones are stored within secretory vesicles. A good example is the endocrine tissue found in the islets of Langerhans in the pancreas. This tissue contains many cell types that are defined by which hormones they produce.
- Secretory vesicles hold the enzymes that are used to make the cell walls of plants, protists, fungi, bacteria, and Archaea cells as well as the extracellular matrix of animal cells.

### Other Types of Vesicles

- Gas vesicles are used by Archaea, bacteria and planktonic microorganisms, possibly to control vertical migration by regulating the gas content and thereby buoyancy, or possibly to position the cell for maximum solar light harvesting.
- Matrix vesicles are located within the extracellular space, or matrix. Using electron microscopy but working independently, they were discovered in 1967 by H. Clarke Anderson and Ermanno Bonucci. These cell-derived vesicles are specialised to initiate biomineralisation of the matrix in a variety of tissues, including bone, cartilage, and dentin. During normal calcification, a major influx of calcium and phosphate ions into the cells accompanies cellular apoptosis (genetically determined self-destruction) and matrix vesicle formation. Calcium-loading also leads to formation of phosphatidylserine:calcium:phosphate complexes in the plasma membrane mediated in part by a protein called annexins. Matrix vesicles bud from the plasma membrane at sites of interaction with the extracellular matrix. Thus, matrix vesicles convey to the extracellular matrix calcium, phosphate, lipids and the annexins which act to nucleate mineral formation. These processes are precisely coordinated to bring about, at the proper place and time, mineralisation of the tissue's matrix unless the Golgi are non-existent.

- Multivesicular body, or MVB, is a membrane-bound vesicle containing a number of smaller vesicles.

### *Vesicle Formation and Transport*

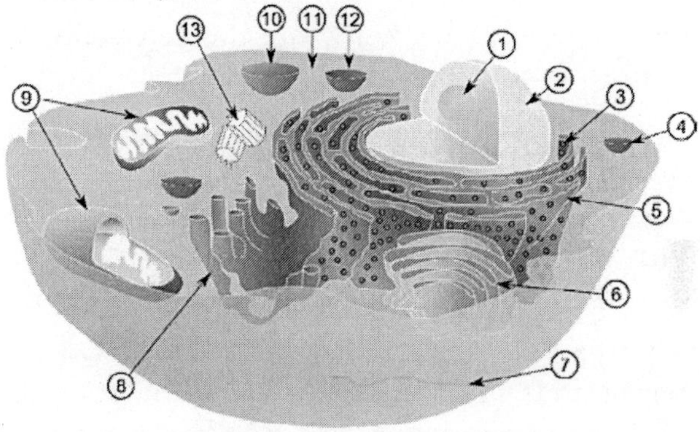


**Figure 23:** Overview of cellular organelles, showing vesicle (4), and related structures (rough endoplasmic reticulum (5) and Golgi apparatus (6).

Some vesicles are made when part of the membrane pinches off the endoplasmic reticulum or the Golgi complex. Others are made when an object outside of the cell is surrounded by the cell membrane.

### *Capturing Cargo Molecules*

The assembly of vesicles requires numerous coats to surround and bind to the proteins being transported. These bind to the coat vesicle. They also trap various transmembrane receptor proteins called cargo receptors, which in turn trap the cargo molecules.

### *Vesicle Coat*

The vesicle coat serves to sculpt the curvature of a donor membrane, and to select specific proteins as cargo. It selects cargo proteins by binding to sorting signals. In this way the vesicle coat clusters selected membrane cargo proteins into nascent vesicle buds.

There are three types of vesicle coats: clathrin, COPI and COPII. Clathrin coats are found on vesicles trafficking between the Golgi and plasma membrane, the Golgi and endosomes, and the plasma membrane and endosomes. COPI coated vesicles are responsible for retrograde transport from the Golgi to the ER, while COPII coated vesicles are responsible for anterograde transport from the ER to the Golgi.

The clathrin coat is thought to assemble in response to regulatory G protein. A coatomer coat assembles and disassembles due to an ARF protein.

### *Vesicle Docking*

Surface markers called SNAREs identify the vesicle's cargo, and complementary SNAREs on the target membrane act to cause fusion of the vesicle and target membrane. Such v-SNARES are hypothesised to exist on the vesicle membrane, while the complementary ones on the target membrane are known as t-SNAREs.

Often SNAREs associated with vesicles or target membranes are instead classified as Qa, Qb, Qc or R SNAREs owing to further variation than simply v- or t-SNAREs. An array of different SNARE complexes can be seen in different tissues and subcellular compartments, with 36 isoforms currently identified in humans.

Regulatory Rab proteins are thought to inspect the joining of the SNAREs. Rab protein is a regulatory GTP-binding protein, and controls the binding of these complementary SNAREs for a long enough time for the Rab protein to hydrolyse its bound GTP and lock the vesicle onto the membrane.

### *Vesicle Fusion*

Fusion requires the two membranes to be brought within 1.5 nm of each other. For this to occur water must be displaced from the surface of the vesicle membrane. This is energetically unfavourable, and evidence suggests that the process requires ATP, GTP and acetyl-coA, fusion is also linked to budding, which is why the term budding and fusing arises.

### *Vesicles in Receptor Downregulation*

Membrane proteins serving as receptors are sometimes tagged for downregulation by the attachment of ubiquitin. After arriving an endosome via the pathway described above, vesicles begin to form inside the endosome, taking with them the membrane proteins meant for degradation; When the endosome either matures to become a lysosome or is united with one, the vesicles are completely degraded. Without this mechanism, only the extracellular part of the membrane proteins would reach the lumen of the lysosome, and only this part would be degraded.

It is because of these vesicles that the endosome is sometimes known as a *multivesicular body*. The pathway to their formation is not completely understood; unlike the other vesicles described above, the outer surface of the vesicles is not in contact with the cytosol.

### Vesicle Preparation

Phospholipid vesicles have been studied in biochemistry. For such studies, a homogeneous phospholipid vesicle suspension can be prepared by sonication, injection of a phospholipid solution into the aqueous buffer solution membranes.

In this way aqueous vesicle solutions can be prepared of different phospholipid composition, as well as different sizes of vesicles.

# Chapter 7

# Supramolecular Metalloproteins

Metalloprotein is a generic term for a protein that contains a metal ion cofactor. Metalloproteins have many different functions in cells, such as enzymes, transport and storage proteins, and signal transduction proteins. Indeed, about one quarter to one third of all proteins require metals to carry out their functions. The metal ion is usually coordinated by nitrogen, oxygen or sulfur atoms belonging to amino acids in the polypeptide chain and/or a macrocyclic ligand incorporated into the protein. The presence of the metal ion allows metalloenzymes to perform functions such as redox reactions that cannot easily be performed by the limited set of functional groups found in amino acids.

## Storage and Transport Metalloproteins

### *Oxygen Carriers*

Hemoglobin, which is the principal oxygen carrier in humans has four sub-units in which the iron(II) ion is coordinated by the planar, macrocyclic ligand protoporphyrin IX (PIX) and the imidazole nitrogen atom of a histidine residue. The sixth coordination site contains a water molecule or a dioxygen molecule. myoglobin has only one such unit. The active site is located in an hydrophobic pocket. This is important as, without it, the iron(II) would be irreversibly oxidised to iron(III). The equilibrium constant for the formation of $HbO_2$ is such that oxygen is taken up or released depending on the partial pressure of oxygen in the lungs or in muscle. In hemoglobin the four sub-units show a cooperativity effect which allows for easy oxygen transfer from hemoglobin to myoglobin. In both hemoglobin and myoglobin it is sometimes incorrectly stated that the oxygenated species contains iron(III). It is now known that the diamagnetic nature of these species is due to the fact that the iron(II) is in the low-spin state. In oxyhemoglobin the iron atom is located in the plane of the porphyrin ring, but in the paramagnetic deoxyhemoglobin the iron atom lies above the plane of the ring. The change in spin state is a cooperative effect of higher crystal field splitting and smaller ionic radius of $Fe^{2+}$ in the oxy- moiety.

Hemerythrin is another iron-containing oxygen carrier. The oxygen binding site is a binuclear iron centre. The iron atoms are coordinated to the protein through the carboxylate side chains of a glutamate and aspartate and five histidine residues. The uptake of $O_2$ by hemerythrin is accompanied by two-electron oxidation of the reduced binuclear centre to produce bound peroxide (OOH-). The mechanism of oxygen uptake and release have been worked out in detail.

Hemocyanins carry oxygen in the blood of most molluscs, and some arthropods such as the horseshoe crab. They are second only to hemoglobin in biological popularity of use in oxygen transport. On oxygenation the two copper(I) atoms at the active site are oxidised to copper(II) and the dioxygen molecules is reduced to peroxide, $O_2^{2-}$

### *Cytochromes*

Oxidation and reduction reactions are not common in organic chemistry as few organic molecules can act as oxidizing or reducing agents. Iron(II), on the other hand, can easily be oxidized to iron(III). This functionality is used in cytochromes which function as electron-transfer vectors. The iron atom in most cytochromes is contained in a heme group. The differences between those cytochromes lies in the different side-chains. For instance Cytochrome a has a heme a prosthetic group and cytochrome b has a heme b prosthetic group. These differences result in different $Fe^{2+}/Fe^{3+}$ redox potentials such that various cytochromes are involved in the mitochondrial electron transport chain.

### *Rubredoxin*

Rubredoxin is an electron-carrier found in sulfur-metabolizing bacteria and archaea. The active site contains an iron ion which is coordinated by the sulphur atoms of four cysteine residues forming an almost regular tetrahedron. Rubredoxins perform one-electron transfer processes. The oxidation state of the iron atom changes between the +2 and +3 states. In both oxidation states the metal is high spin, which helps to minimize structural changes.

## Iron-sulfur Protein

Iron-sulfur proteins are proteins characterized by the presence of iron-sulfur clusters containing sulfide-linked di-, tri-, and tetrairon centres in variable oxidation states. Iron-sulfur clusters are found in a variety of metalloproteins, such as the ferredoxins, as well as NADH dehydrogenase, hydrogenases, Coenzyme Q - cytochrome c reductase, Succinate- coenzyme Q reductase and nitrogenase. Iron-sulfur clusters are best known for their role in the oxidation-reduction reactions of mitochondrial electron transport. Both Complex I and Complex II of oxidative phosphorylation have multiple Fe-S clusters. They have many other functions including catalysis as illustrated by aconitase, generation of radicals as illustrated by SAM-dependent enzymes, and as sulfur donors in the biosynthesis of lipoic acid and biotin. Additionally some Fe-S proteins regulate gene expression. Fe-S proteins are vulnerable to attack by biogenic nitric oxide.

## Structural Motifs

In almost all Fe-S proteins, the Fe centres are tetrahedral and the terminal ligands are thiolato sulfur centres from cysteinyl residues. The sulfide groups are either two- or three-coordinated. Three distinct kinds of Fe-S clusters with these features are most common.

### *2Fe-2S Clusters*

The simplest polymetallic system, the $[Fe_2S_2]$ cluster, is constituted by two iron ions bridged by two sulfide ions and coordinated by four cysteinyl ligands (in $Fe_2S_2$ ferredoxins) or by two cysteines and two histidines (in Rieske proteins). The oxidized proteins contain two $Fe^{3+}$ ions, whereas the reduced proteins contain one $Fe^{3+}$ and one $Fe^{2+}$ ion. These species exist in two oxidation states, $(Fe^{III})_2$ and $Fe^{III}Fe^{II}$.

### *4Fe-4S Clusters*

A common motif features a four iron ions and four sulfide ions placed at the vertices of a cubane-type structure. The Fe centres are typically further coordinated by cysteinyl ligands. The $[Fe_4S_4]$ electron-transfer proteins ($[Fe_4S_4]$ ferredoxins) may be further subdivided into low-potential (bacterial-type) and high-potential (HiPIP) ferredoxins.

In HiPIP, the cluster shuttles between $[2Fe^{3+}, 2Fe^{2+}]$ $(Fe_4S_4^{2+})$ and $[3Fe^{3+}, Fe^{2+}]$ $(Fe_4S_4^{3+})$. The potentials for this redox couple range from 0.4 to 0.1 V. In the bacterial ferredoxins, the pair of oxidation states are $[Fe^{3+}, 3Fe^{2+}]$ $(Fe_4S_4^{+})$ and $[2Fe^{3+}, 2Fe^{2+}]$ $(Fe_4S_4^{2+})$. The potentials for this redox couple range from -0.3 to -0.7 V. The two families of 4Fe-4S clusters share the $Fe_4S_4^{2+}$ oxidation state. The difference in the redox couples is attributed to the degree of hydrogen bonding, which strongly modifies the basicity of the cysteinyl thiolate ligands. A further redox couple, which is still more reducing than the bacterial ferredoxins is implicated in the nitrogenase.

Some 4Fe-4S clusters bind substrates and are thus classified as enzymes. In aconitase, the Fe-S cluster binds aconitate at the one Fe centre that lacks a thiolate ligand. The cluster does not undergo redox, but serves as a Lewis acid catalyst to convert aconitate to isocitrate. In the radical-SAM enzymes, the cluster binds and reduces S-adenosylmethionine to generate a radical, which is involved in many biosyntheses.

### 3Fe-4S Clusters

Proteins are also known to contain $[Fe_3S_4]$ centres, which feature one iron less than the more common $[Fe_4S_4]$ cores. Three sulfide ions bridge two iron ions each, while the fourth sulfide bridges three iron ions. Their formal oxidation states may vary from $[Fe_3S_4]$ (all-$Fe^{3+}$ form) to $[Fe_3S_4]^{2-}$ (all-$Fe^{2+}$ form). In a number of iron-sulfur proteins, the $[Fe_4S_4]$ cluster can be reversibly converted by oxidation and loss of one iron ion to a $[Fe_3S_4]$ cluster. E.g., the inactive form of aconitase possesses an $[Fe_3S_4]$ and is activated by addition of $Fe^{2+}$ and reductant.

### Other Fe-S Clusters

More complex polymetallic systems are common. Examples include both the 8Fe and the 7Fe clusters in nitrogenase. Carbon monooxide dehydrogenase and the [FeFe]-hydrogenase also feature unusual Fe-S clusters.

### Biosynthesis

The biosynthesis of the Fe-S clusters has been well studied. The biogenesis of iron sulfur clusters has been studied most extensively in the bacteria *E. coli* and *A. vinelandii* and yeast *S. cerevisiae.* At least three different biosynthetic systems have been identified so far, namely nif, suf, and isc systems, which were first identified in bacteria.

The nif system is responsible the clusters in the enzyme nitrogenase. The suf and isc systems are more general with the isc-related proteins being present only in the animal kingdom. The yeast isc system is the best described.

Several proteins constitute the biosynthetic machinery via the isc pathway. The process occurs in two major steps: (1) the Fe/S cluster is assembled on a scaffold protein followed by (2) transfer of the preformed cluster to the recipient proteins. The first step of this process occurs in the cytoplasm of prokaryotic organisms or in the mitochondria of eukaryotic organisms. In the higher organisms the clusters are therefore transported out of the mitochondrion to be incorporated into the extramitochondrial enzymes.

These organisms also possess a set of proteins involved in the Fe/S clusters transport and incorporation processes that are not homologous to proteins found in procaryotic systems.

### *Synthetic Analogues*

Synthetic analogues of the naturally occurring Fe-S clusters were first reported by Holm and coworkers. Treatment of iron salts with a mixture of thiolates and sulfide affords derivatives such as $(Et_4N)_2Fe_4S_4(SCH_2Ph)_4]$.

### *Plastocyanin*

Plastocyan is one of the family of blue copper proteins which are involved in electron transfer reactions. The copper binding site is described as a 'distorted trigonal pyramidal'. The trigonal plane of the pyramidal base is composed of two nitrogen atoms ($N_1$ and $N_2$) from separate histidines and a sulfur ($S_1$) from a cysteine. Sulfur ($S_2$) from an axial methionine forms the apex.

The 'distortion' occurs in the bond lengths between the copper and sulfur ligands. The Cu-$S_1$ contact is shorter (207 picometers) than Cu-$S_2$ (282 pm). The elongated Cu-$S_2$ bonding destabilises the $Cu^{II}$ form and increases the redox potential of the protein. The blue colour (597 nm peak absorption) is due to the Cu-$S_1$ bond where $S_{p\eth}$ to $Cu_{dx2-y2}$ charge transfer occurs.

In the reduced form of plastocyanin, His-87 will become protonated with a $pK_a$ of 4.4. Protonation prevents it acting as a ligand and the copper site geometry becomes trigonal planar.

### *Iron Storage and Transfer*

Iron is stored as iron(III) in ferritin. The exact nature of the binding site has not yet been determined. The iron appears to be present as an hydrolysis product such as FeO(OH). Iron is transported by transferrin whose binding site consists of two tyrosines, one aspartic acid and one histidine.

The human body has no mechanism for iron excretion. This can lead to iron-overload problems in patients treated with blood transfusions, as, for instance, with β-thallasemia.

### Ceruloplasmin

Ceruloplasmin is the major copper-carrying protein in the blood. Ceruloplasmin exhibits oxidase activity, which is associated with possible oxidation of $Fe^{2+}$ (ferrous iron) into $Fe^{3+}$ (ferric iron), therefore assisting in its transport in the plasma in association with transferrin, which can only carry iron in the ferric state.

### Metalloenzymes

Metalloenzymes all have one feature in common, namely, that the metal ion is bound to the protein with one labile coordination site. As with all enzymes, the shape of the active site is crucial. The metal ion is usually located in a pocket whose shape fits the substrate. The metal ion catalyzes reactions which are difficult to achieve in organic chemistry.

### Carbonic Anhydrase

This reaction is very slow in the absence of a catalyst, but quite fast in the presence of the hydroxide ion

$$CO_2 + OH^- \rightleftharpoons HCO_3^-$$

A reaction similar to this is almost instantaneaous with carbonic anhydrase. The structure of the active site in carbonic anhydrases is well-known from a number of crystal structures. It consists of a zinc ion coordinated by three imidazole nitrogen atoms from three histidine units.

The fourth coordination site is occupied by a water molecule. The coordination sphere of the zinc ion is approximately tetrahedral.

The positively charged zinc ion polarizes the coordinated water molecule and nucleophilic attack by the negatively charged hydoxide portion on carbon dioxide (carbonic anhydride) proceeds rapidly. The catalytic cycle produces the bicarbonate ion and the hydrogen ion as the equilibrium

$$H_2CO_3 \; HCO_3^- + H^+$$

Favours dissociation of carbonic acid at biological pH values.

### Vitamin $B_{12}$-dependent Enzymes

Vitamin $B_{12}$ catalyzes the transfer of methyl ($-CH_3$) groups between two molecules, which involves the breaking of C-C bonds, a process that is energetically expensive in organic reactions. The metal ion lowers the activation energy for the process by forming a transient Co-$CH_3$ bond. The structure of the coenzyme was famously determined by Dorothy Hodgkin and co-workers, for which she received a Nobel prize. It constists of a cobalt(II) ion coordinated by four nitrogen atoms of a corrin rings and a fifth Nitrogen atom from an imidazole group. In the resting state there is a Co—C σ bond with the 5' carbon atom of adenosine. This is a naturally occurring organometallic compound, which explains its function in trans-methylation reactions, such as the reaction carried out by methionine synthase.

### *Nitrogenase (Nitrogen Fixation)*

The fixation of atmospheric nitrogen is a very energy-intensive process, as it involves breaking the very stable triple bond between the nitrogen atoms. The enzyme nitrogenase is one of the few enzymes that can catalyse the process. The enzyme occurs in certain bacteria.

There are three components to its action: a molybdenum atom at the active site, Iron-sulfur clusters which are involved in transporting the electrons needed to reduce the nitrogen and an abundant energy source. The energy is provided by a symbiotic relationship between the bacteria and a host plant, often a legume. The relationship is symbiotic because the plant supplies the energy by photosynthesis and benefits by obtaining the fixed nitrogen. The reaction may be written symbolically as

$$N_2 + 16\ MgATP + 8e \rightarrow 2NH_3 + 16\ MgADP + 16\ P_i + H_2$$

Where $P_i$ stands for inorganic phosphate. The precise structure of the active site has been difficult to determine. It appears to contain a $MoFe_7S_8$ cluster which is able to bind the dinitrogen molecule and, presumably, enable the reduction process to begin. The electrons are transported by the associated "P" cluster, which contains two cubical $Fe_4S_4$ clusters joined by sulphur bridges.

### *Superoxide Dismutase*

The superoxide ion, $O_2^-$ is generated in biological systems by reduction of molecular oxygen. It has an unpaired electron, so it behaves as a free radical. It is a powerful oxidising agent. These properties render the superoxide ion very toxic and are deployed to advantage by phagocytes to kill invading micro organisms. Otherwise, the superoxide ion must be destroyed before it does unwanted damage in a cell. The superoxide dismutase enzymes perform this function very efficiently.

The formal oxidation state of the oxygen atoms is ½. In solutions at neutral pH, the superoxide ion disproportionates to molecular oxygen and hydrogen peroxide.

$$2\ O_2^- + 2\ H^+ \rightarrow O_2 + H_2O_2$$

In biology this type of reaction is call a dismutation reaction. It involves both oxidation and reduction of superoxide ions. The superoxide dismutase group of enzymes, abbreviated as SOD, increase the rate of reaction to near the diffusion limited rate. The key to the action of these enzymes is a metal ion with variable oxidation state which can act as either an oxidizing agent or as a reducing agent.

$$\text{Oxidation: } M^{(n+1)+} + O_2^- \rightarrow M^{n+} + O_2$$

$$\text{Reduction: } M^{n+} + O_2^- + 2H^+ \rightarrow M^{(n+1)+} + H_2O_2.$$

In human SOD the active metal is copper, as $Cu^{2+}$ or $Cu^+$, coordinated tetrahedrally by four histidine residues. This enzyme also contains zinc ions. Other isozymes may contain iron, manganese or nickel.

Ni-SOD is particularly interesting as it involves nickel(III), an unusual oxidation state for this element. The active site Ni geometry cycles from square planar Ni(II), with thiolate (Cys2 and Cys6) and backbone nitrogen (His1 and Cys2) ligands, to square pyramidal Ni(III) with an added axial His1 side chain ligand.

### *Chlorophyll-containing Proteins*

Chlorophyll plays a crucial role in photosynthesis. It contains a magnesium enclosed in a chlorin ring. However, the magnesium ion is not directly involved in the photosynthetic function and can be replaced by other divalent ions with little loss of activity. Rather, the photon is absorbed by the chlorin ring, whose electronic structure is well-adapted for its purpose.

Initially, the absorption of a photon causes an electron to be excited into a singlet state of the Q band. The excited state undergoes an intersystem crossing from the singlet state to a triplet state in which there are two electrons with parallel spin. This species is, effectively, a free radical, and is very reactive and allows an electron to be transferred to acceptors which are adjacent to the chlorophyll in the chloroplast.

In the process chlorophyll is oxidized. Later in the phosynthetic cycle, chlorophyll is re-reduced. This reduction ultimately draws electrons from water, yielding molecular oxygen as a final oxidation product.

## Signal-transduction Metalloproteins

### Calmodulin

Calmodulin is an example of a signal-transduction protein. It is a small protein which contains four EF-hand motifs, each of which can bind a $Ca^{2+}$ ion. In an EF-hand loop the calcium ion is coordinated in a pentagonal bipyramidal configuration. Six Glutamic acid and Aspartic acid residues involved in the binding are in positions 1, 3, 5, 7, 9 of the polypeptide chain.

At position 12 there is a glutamate or aspartate ligand which behaves as a (bidentate ligand), providing two oxygen atoms. The ninth residue in the loop is necessarily glycine due to the conformational requirements of the backbone. The coordination sphere of the calcium ion contains only carboxylate oxygen atoms and no nitrogen atoms. This is consistent with the hard nature of the calcium ion.

The protein has two approximately symmetrical domains, separated by a flexible "hinge" region. Binding of calcium causes a conformational change to occur in the protein. Calmodulin participates in an intracellular signalling system by acting as a diffusible second messenger to the initial stimuli.

### Transcription Factors

Many transcription factors contain a structure known as a zinc finger, this is a structural module where a region of protein folds around a zinc ion. The zinc does not directly contact the DNA that these proteins bind to, instead the cofactor is essential for the stability of the tightly-folded protein chain. In these proteins the zinc ion is usually coordinated by pairs of cysteine and histidine side chains.

### Other Metalloenzymes

| *Ion* | *Examples of enzymes containing this ion* |
|---|---|
| Magnesium | Glucose 6-phosphataseHexokinaseDNA polymerase |
| Vanadium | vanabins |
| Manganese | Arginase |
| Iron | CatalaseHydrogenaseIRE-BPAconitase |
| Nickel | UreaseHydrogenase |
| Copper | Cytochrome oxidaseLaccase |
| Zinc | Alcohol dehydrogenase Carboxypeptidase Amino-peptidase Beta amyloid |
| Molybdenum | Nitrate reductase |
| Selenium | Glutathione peroxidase |
| various | MetallothioneinPhosphatase |

In metalloproteins, the properties of one or more metal ions are tuned by a protein environment to produce efficient and selective catalysts for a large variety of reactions, e.g. the activation of small inorganic molecules like oxygen and nitrogen, group transfer, redox reactions, and solvolysis.

Bioinorganic chemists study the characterization of the function and structure of metalloproteins and structural and functional models for them. Supramolecular chemists,293 studying the chemistry of noncovalent bonds in areas like host-guest chemistry or the chemistry of self-assembling molecules, are showing an increasing interest in the development of metalloprotein mimics that combine a metal in a catalytic site with a hydrophobic binding pocket like a cavity or a membrane. In particular catalysts for the oxidation of hydrocarbons and sensor systems whose redox potentials depend on the concentration of alkali cations, we have designed and synthesized a number of selfassembling and cavitand systems containing metal clusters.

As part of our programme aimed at the design and synthesis of new supramolecular systems,

## CYTOCHROME P-450

Cytochrome P-4505 is a monooxygenase, i.e. it catalyses the reductive activation of molecular oxygen. One oxygen atom is incorporated in an organic substrate, and the other is reduced to water. The electrons required for the reaction are obtained from the nicotinamide redox cofactor NADPH + H+ via the flavoprotein NADPH-cytochrome P-450 reductase. In the proposed catalytic cycle,6 the electrons are transferred separately, viz. after the binding of the hydrocarbon substrate and oxygen, respectively. The crystal structure7 of cytochrome P-450 shows that there is a substrate binding cavity close to a catalytic iron porphyrin, which has a thiolate as an axial ligand. Cytochrome P-450 'mimics' possess Mnporphyrins and Fe-porphyrins as metal complexes, and often catalyse oxygen transfer from a single oxygen donor like iodosobenzene instead of molecular 0xygen.An interesting aspect of cytochrome P-450 is that it is a membrane-bound enzyme, and that phospholipids stimulate the transfer of electrons to the isolated enzyme and enhance its affinity for the substrate.

This feature has stimulated a number of groupsll-13 to develop cytochrome P-450 mimics in which a hydrophobic porphyrin is incorporated within the bilayer of a vesicle membrane.

In our first model, the electrons required for the partial reduction of oxygen were derived from the oxidation of molecular hydrogen by colloidal Pt.14 This approach has the disadvantage, apart from the potentially hazardous combination of hydrogen and oxygen, that a non-productive reaction between these gases is possible, viz. water formation. In a later stage in the development of our system 15 we applied another reaction, viz. the Rh-catalysed oxidation of formate to carbon dioxide,l6 for the supply of the reducing equivalents.

In order to determine which porphyrin would be most suited as a catalytic group in our cytochrome P-450 mimic, the location, aggregation and orientation of a number of tetraarylporphyrins in vesicle bilayers of both positively and negatively charged amphiphiles, viz. dioctadecyl-dimethylammonium chloride (DODAC) and sodium dihexylphosphate (DHP), respectively, were investigated in detail. A study of the effects of hydrophobic and hydrophilic quenchers on the porphyrin fluorescence in DODAC bilayers revealed that the hydrophobic porphyrin, 2 1, 23-2H-5, 10, 15, 20-tetra (4-hexadecyloxypheny1) porphyrin (THPP, 1) is located in the centre of the bilayer, whereas the positively charged porphyrin 2 1, 23-2H-5-(1 -methyl-4-pyridyl)- 10, 15, 20-tri (4-hexadecyloxyphenyl) porphyrin monotosylate (2) is close to the surface of the vesicles. The aggregation behaviour of the porphyrins was investigated by looking at the self-quenching of their fluorescence as a function of the [amphiphile]/[porphyrin] ratio. Following the series 1-5, the aggregation in DHP vesicles decreases as the number of positive charges on the porphyrin increases, which can be explained by an increasing electrostatic repulsion between the porphyrin molecules.

In contrast, the aggregation of the porphyrins in the series 2-4 in DODAC bilayers was found to increase markedly with the number of positive charges, and 1, 23-2H-5, 10,15, 20-tetr1a- (m ethyl-4-pyridyl)-porphyrin tetratosylate (5) could not be incorporated in DODAC bilayers. These results indicate that in this case the electrostatic repulsion between porphyrin and amphiphile is the dominant factor. Steric effects are probably even more important than electrostatic effects, since the association constant of o-dichlorosubstituted tetraphenylporphyrin (TDCPP, 7) in DODAC bilayers was found to be lower (KSs = 6.5 x 103 M-1) than that of the other hydrophobic porphyrins studied, viz. tetraphenylporphyrin (TPP, 6) and THPP (l),as well as the charged porphyrins *24* (typical Kass = 1 to 4 x 104 M-I). The aggregation-induced shifts in the UV-vis spectra of the porphyrins could be interpreted according to the exciton model. We concluded that in DODAC bilayers, the charged porphyrins 2, 4 form face-to-face aggregates, whereas the hydrophobic porphyrin 6 aggregates in an edge-to-edge fashion.

The orientation of the porphyrins in the bilayers was investigated by measuring the EPR spectra of the porphyrins in dried bilayers at various orientations. 18919 It was established that monomers of 7 have an orientation perpendicular to the normal of the DODAC bilayer. We decided to use porphyrin 7 as the catalytic group in our cytochrome P-450 model, because its aggregation appears to be prohibited by the bulky o-substituents and therefore its catalytic efficiency should not be impaired.

These compounds are generally produced by the self-assembly of small molecules held together by non-covalent interactions (hydrogen bonds, or weak intermolecular interactions). They allow molecular recognition. The example below is redraw from a article published recently. These ions are trivalent lanthanum, $La^{3+}$ and trivalent europium $Eu^{3+}$; they require nine chemical bonds each, which is achieved by combining three ligand strands (the tailored molecule) with a mixture of $La^{3+}$ and $Eu^{3+}$ ions.

## Molecular Binding of Metal Complexes by Macroycles

Molecular binding is an attractive interaction between two molecules which results in a stable association in which the molecules are in close proximity to each other. The result of molecular binding is formation of a molecular complex.

## Types

Molecular binding can be classified into the following types:

- non-covalent – no chemical bonds are formed between the two interacting molecules hence the association is fully reversible
- reversible covalent – a chemical bond is formed, however the free energy difference separating the non-bonded reactants from bonded product is near equilibrium and the activation barrier is relatively low such that the reverse reaction which cleaves the chemical bond easily occurs
- irreversible covalent – a chemical bond is formed in which the product is thermodynamically much more stable than the reactants such that the reverse reaction does not take place.

### *Examples*

Molecules that can participate in molecular binding include proteins, nucleic acids, carbohydrates, lipids, and small organic molecules such as drugs. Hence the types of complexes that form as a result of molecular binding include:

- protein – protein
- protein – DNA
- protein – hormone
- protein – drug.

Proteins that form stable complexes with other molecules are often referred to as receptors while their binding partners are called ligands.

In chemistry, a coordination complex or metal complex, is a structure consisting of a central atom or ion (usually metallic), bonded to a surrounding array of molecules or anions (ligands, complexing agents). The atom within a ligand that is directly bonded to the central atom or ion is called the donor atom. Polydentate (multiple bonded) ligands can form a chelate complex. A ligand donates at least one pair of electrons to the central atom/ion.

Compounds that contain a coordination complex are called coordination compounds. The central atom or ion, together with all ligands form the coordination sphere.

Coordination refers to the "coordinate covalent bonds" (dipolar bonds) between the ligands and the central atom. Originally, a complex implied a reversible association of molecules, atoms, or ions through such weak chemical bonds. As applied to coordination chemistry, this meaning has evolved. Some metal complexes are formed virtually irreversibly and many are bound together by bonds that are quite strong.

Coordination complexes were known - although not understood in any sense - since the beginning of chemistry, e.g. Prussian blue and copper vitriol. The key breakthrough occurred when Alfred Werner proposed in 1893 that Co(III) bears six ligands in an octahedral geometry. His theory allows one to understand the difference between coordinated and ionic in a compound, for example chloride in the cobalt ammine chlorides and to explain many of the previously inexplicable isomers.

In 1914, Werner resolved the first coordination complex, called hexol, into optical isomers, overthrowing the theory that chirality was necessarily associated with carbon compounds.

### *Structure of Coordination Complexes*

The ions or molecules surrounding the central atom are called ligands. Ligands are generally bound to the central atom by a coordinate covalent bond (donating electrons from a lone electron pair into an empty metal orbital), and are said to be coordinated to the atom. There are also organic ligands such as alkenes whose pi bonds can coordinate to empty metal orbitals. An example is ethene in the complex known as Zeise's salt, $K^+[PtCl_3(C_2H_4)]^-$.

### *Geometry*

In coordination chemistry, a structure is first described by its coordination number, the number of ligands attached to the metal (more specifically, the number of σ-type bonds between ligand(s) and the central atom). Usually one can count the ligands attached, but sometimes even the counting can become ambiguous. Coordination numbers are normally between two and nine, but large numbers of ligands are not uncommon for the lanthanides and actinides. The number of bonds depends on the size, charge, and electron configuration of the metal ion and the ligands. Metal ions may have more than one coordination number.

Typically the chemistry of complexes is dominated by interactions between s and p molecular orbitals of the ligands and the d orbitals of the metal ions. The s, p, and d orbitals of the metal can accommodate 18 electrons. The maximum coordination number for a certain metal is thus related to the electronic configuration of the metal ion (more specifically, the number of empty orbitals) and to the ratio of the size of the ligands and the metal ion. Large metals and small ligands lead to high coordination numbers, e.g. $[Mo(CN)_8]^{4-}$. Small metals with large ligands lead to low coordination numbers, e.g. $Pt[P(CMe_3)]_2$. Due to their large size, lanthanides, actinides, and early transition metals tend to have high coordination numbers.

Different ligand structural arrangements result from the coordination number. Most structures follow the points-on-a-sphere pattern (or, as if the central atom were in the middle of a polyhedron where the corners of that shape are the locations of the ligands), where orbital overlap (between ligand and metal orbitals) and ligand-ligand repulsions tend to lead to certain regular geometries. The most observed geometries are listed below, but there are many cases which deviate from a regular geometry, e.g. due to the use of ligands of different types (which results in irregular bond lengths; the coordination atoms do not follow a points-on-a-sphere pattern), due to the size of ligands, or due to electronic effects:

- Linear for two-coordination,
- Trigonal planar for three-coordination,
- Tetrahedral or square planar for four-coordination
- Trigonal bipyramidal or square pyramidal for five-coordination,
- Octahedral (orthogonal) or trigonal prismatic for six-coordination,
- Pentagonal bipyramidal for seven-coordination,
- Square antiprismatic for eight-coordination, and
- Tri-capped trigonal prismatic (Triaugmented triangular prism) for nine coordination.

Some exceptions and provisions should be noted:

- The idealized descriptions of 5-, 7-, 8-, and 9- coordination are often indistinct geometrically from alternative structures with slightly different L-M-L (ligand-metal-ligand) angles. The classic example of this is the difference between square pyramidal and trigonal bipyramidal structures.
- Due to special electronic effects such as (second-order) Jahn-Teller stabilization, certain geometries are stabilized relative to the other possibilities, e.g. for some compounds the trigonal prismatic geometry is stabilized relative to octahedral structures for six-coordination.

### *Isomerism*

The arrangement of the ligands is fixed for a given complex, but in some cases it is mutable by a reaction that forms another stable isomer. There exist many kinds of isomerism in coordination complexes, just as in many other compounds.

### *Stereoisomerism*

Stereoisomerism occurs with the same bonds in different orientations relative to one another. Stereoisomerism can be further classified into:

### *Cis-trans Isomerism and Facial-meridional Isomerism*

Cis-trans isomerism occurs in octahedral and square planar complexes (but not tetrahedral). When two ligands are mutually adjacent they are said to be cis, when opposite each other, trans. When three identical ligands occupy one face of an octahedron, the isomer is said to be facial, or fac. In a fac isomer, any two identical ligands are adjacent or cis to each other. If these three ligands and the metal ion are in one plane, the isomer is said to be meridional, or mer.

### Optical Isomerism

Optical isomerism occurs when the mirror image of a compound is not superimposable with the original compound. It is so called because such isomers are optically active, that is, they rotate the plane of polarized light. The symbol Λ (*lambda*) is used as a prefix to describe the left-handed propeller twist formed by three bidentate ligands.

### Structural Isomerism

Structural isomerism occurs when the bonds are themselves different. Linkage isomerism is only one of several types of structural isomerism in coordination complexes (as well as other classes of chemical compounds). Linkage isomerism occurs with ambidentate ligands which can bind in more than one place. For example, $NO_2$ is an ambidentate ligand: it can bind to a metal at either the N atom or at an O atom.

### Electronic Properties of Metal Complexes

Many of the properties of metal complexes are dictated by their electronic structures. The electronic structure can be described by a relatively ionic model that ascribes formal charges to the metals and ligands. This approach is the essence of crystal field theory (CFT). Crystal field theory, introduced by Hans Bethe in 1929, gives a quantum mechanically based attempt at understanding complexes. But crystal field theory treats all interactions in a complex as ionic and assumes that the ligands can be approximated by negative point charges.

More sophisticated models embrace covalency, and this approach is described by ligand field theory (LFT) and Molecular orbital theory (MO). Ligand field theory, introduced in 1935 and built from molecular orbital theory, can handle a broader range of complexes and can explain complexes in which the interactions are covalent. The chemical applications of group theory can aid in the understanding of crystal or ligand field theory, by allowing simple, symmetry based solutions to the formal equations.

Chemists tend to employ the simplest model required to predict the properties of interest; for this reason, CFT has been a favourite for the discussions when possible. MO and LF theories are more complicated, but provide a more realistic perspective.

The electronic configuration of the complexes gives them some important properties:

### Colour

Metal complexes often have spectacular colours caused by electronic transitions by the absorption of light. For this reason they are often applied as pigments. Most transitions that are related to coloured metal complexes are either d-d transitions or charge transfer bands. In a d-d transition, an electron in a d orbital on the metal is excited by a photon to another d orbital of higher energy. A charge transfer band entails promotion of an electron from a metal-based orbital into an empty ligand-based orbital (Metal-to-Ligand Charge Transfer or MLCT). The converse also occurs: excitation of an electron in a ligand-based orbital into an empty metal-based orbital (Ligand to Metal Charge Transfer or LMCT). These phenomena can be observed with the aid of electronic spectroscopy; also known as UV-Vis. For simple compounds with high symmetry, the d-d transitions can be assigned using Tanabe-Sugano diagrams. Increasingly, these assignments can be supported with computational chemistry.

### Magnetism

Metal complexes that have unpaired electrons are magnetic. Considering only monometallic complexes, unpaired electrons arise because the complex has an odd number of electrons or because electron pairing is destabilized. Thus, monomeric Ti(III) species have one "d-electron" and must be (para)magnetic, regardless of the geometry or the nature of the ligands. Ti(II), with two d-electrons, forms some complexes that have two unpaired electrons and others with none. This effect is illustrated by the compounds $TiX_2[(CH_3)_2PCH_2CH_2P(CH_3)_2]_2$: when X = Cl, the complex is paramagnetic (high spin configuration) whereas when $X=CH_3$, it is diamagnetic (low spin configuration). It is important to realize that ligands provide an important means of adjusting the ground state properties.

In bi- and polymetallic complexes, in which the individual centres have an odd number of electrons or which are high spin, the situation is more complicated. If there is interaction (either direct or through ligand) between the two (or more) metal centres, the electrons may couple (antiferromagnetic coupling, resulting in a diamagnetic compound), or they may enhance each other (ferromagnetic coupling).

When there is no interaction, the two (or more) individual metal centres behave as if in two separate molecules.

### Reactivity

Complexes show a variety of possible reactivities:

***Electron transfers :*** A common reaction between coordination complexes involving ligands are inner and outer sphere electron transfers. They are two different mechanisms of electron transfer redox reactions, largely defined by the late Henry Taube. In an inner sphere reaction, a ligand with two lone electron pairs acts as a *bridging ligand*, a ligand to which both coordination centres can bond. Through this, electrons are transferred from one centre to another.

***(Degenerate) ligand exchange :*** One important indicator of reactivity is the rate of degenerate exchange of ligands. For example, the rate of interchange of coordinate water in $[M(H_2O)_6]^{n+}$ complexes varies over 20 orders of magnitude. Complexes where the ligands are released and rebound rapidly are classified as labile. Such labile complexes can be quite stable thermodynamically. Typical labile metal complexes have either low charge ($Na^+$), electrons in d-orbitals that are antibonding with respect to the ligands ($Zn^{2+}$), or lack covalency ($Ln^{3+}$, where Ln is any lanthanide). The lability of a metal complex also depends on the high vs. low spin configurations when such is possible. Thus high spin Fe(II) and Co(III) form labile complexes whereas low spin analogues are inert. Cr(III) can only exist in the low-spin state (quartet), which is inert because of its high formal oxidation state, absence of electrons in orbitals that are M-L antibonding, plus some “ligand field stabilization” associated with the d3 configuration.

***Associative processes :*** Complexes that have unfilled or half-filled orbitals often show the capability to react with substrates. Most substrates have a singlet ground-state; that is, they have lone electron pairs (e.g. water, amines, ethers), so these substrates need an empty orbital to be able to react with a metal centre. Some substrates (e.g. molecular oxygen) have a triplet ground state, which results that metals with half-filled orbitals have a tendency to react with such substrates (it must be said that the dioxygen-molecule also has lone pairs, so it is also capable to react as a ‘normal’ Lewis base).

If the ligands around the metal are carefully chosen, the metal can aid in (stoichiometric or catalytic) transformations of molecules or be used as a sensor.

### *Classification*

Metal complexes, also known as coordination compounds, include all metal compounds, aside from metal vapors, plasmas, and alloys. The study of "coordination chemistry" is the study of "inorganic chemistry" of all alkali and alkaline earth metals, transition metals, lanthanides, actinides, and metalloids. Thus, coordination chemistry is the chemistry of the majority of the periodic table. Metals and metal ions only exist, in the condensed phases at least, surrounded by ligands.

The areas of coordination chemistry can be classified according to the nature of the ligands, broadly speaking:

- Classical (or "Werner Complexes"): Ligands in classical coordination chemistry bind to metals, almost exclusively, via their "lone pairs" of electrons residing on the main group atoms of the ligand. Typical ligands are $H_2O$, $NH_3$, $Cl^-$, $CN^-$, $en^-$

Examples: $[Co(EDTA)]^-$, $[Co(NH_3)_6]Cl_3$, $[Fe(C_2O_4)_3]K_3$

- Organometallic Chemistry: Ligands are organic (alkenes, alkynes, alkyls) as well as "organic-like" ligands such as phosphines, hydride, and CO.

Example: $(C_5H_5)Fe(CO)_2CH_3$

- Bioinorganic Chemistry: Ligands are those provided by nature, especially including the side chains of amino acids, and many cofactors such as porphyrins.

Example: hemoglobin;

Many natural ligands are "classical" especially including water.

- Cluster Chemistry: Ligands are all of the above also include other metals as ligands.

Example $Ru_3(CO)_{12}$

- In some cases there are combinations of different fields:

Example: $[Fe_4S_4(Scysteinyl)_4]^{2-}$, in which a cluster is embedded in a biologically active species.

Mineralogy, materials science, and solid state chemistry - as they apply to metal ions - are subsets of coordination chemistry in the sense that the metals are surrounded by ligands. In many cases these ligands are oxides or sulfides, but the metals are coordinated nonetheless, and the principles and guidelines. In hydrates, at least some of the ligands are water molecules. It is true that the focus of mineralogy, materials science, and solid state chemistry differs from the usual focus of coordination or inorganic chemistry. The former are primarily concerned with polymeric structures, properties arising from a collective effects of many highly interconnected metals.

In contrast, coordination chemistry focuses on reactivity and properties of complexes containing individual metal atoms or small ensembles of metal atoms.

### *Older Classifications of Isomerism*

Traditional classifications of the kinds of isomer have become archaic with the advent of modern structural chemistry. In the older literature, one encounters:

- Ionisation isomerism describes the possible isomers arising from the exchange between the outer sphere and inner sphere. This classification relies on an archaic classification of the inner and outer sphere. In this classification, the "outer sphere ligands," when ions in solution, may be switched with "inner sphere ligands" to produce an isomer.
- Solvation isomerism occurs when an inner sphere ligand is replaced by a solvent molecule. This classification is obsolete because it considers solvents as being distinct from other ligands. Some of the problems are discussed under water of crystallization.

### *Naming Complexes*

The basic procedure for naming a complex:

1. When naming a complex ion, the ligands are named before the metal ion.
2. Write the names of the ligands in alphabetical order. (Numerical prefixes do not affect the order.)
   - Multiple occurring monodentate ligands receive a prefix according to the number of occurrences: *di-*, *tri-*, *tetra-*, *penta-*, or *hexa*. Polydentate ligands (e.g., ethylenediamine, oxalate) receive *bis-*, *tris-*, *tetrakis-*, etc.
   - Anions end in *ido*. This replaces the final 'e' when the anion ends with '-ate', e.g. *sulfate* becomes *sulfato*. It replaces 'ide': *cyanide* becomes *cyanido*.
   - Neutral ligands are given their usual name, with some exceptions: $NH_3$ becomes *ammine*; $H_2O$ becomes *aqua* or *aquo*; CO becomes *carbonyl*; NO becomes *nitrosyl*.

3. Write the name of the central atom/ion. If the complex is an anion, the central atom's name will end in *-ate*, and its Latin name will be used if available (except for mercury).
4. If the central atom's oxidation state needs to be specified (when it is one of several possible, or zero), write it as a Roman numeral (or 0) in parentheses.
5. Name cation then anion as separate words (if applicable, as in last example).

Examples:

$[NiCl_4]^{2-}$ → tetrachloridonickelate(II) ion

$[CuNH_3Cl_5]^{3-}$ → amminepentachloridocuprate(II) ion

$[Cd(en)_2(CN)_2]$ →

dicyanidobis(ethylenediamine)cadmium(II)

$[Co(NH_3)_5Cl]SO_4$ → pentaamminechloridocobalt(III) sulfate

The coordination number of ligands attached to more than one metal (bridging ligands) is indicated by a subscript to the Greek symbol μ placed before the ligand name. Thus the dimer of aluminium trichloride is described by $Al_2Cl_4(\mu_2\text{-}Cl)_2$.

### *Application of Coordination Compounds*

(1) They are used in photography, i.e., AgBr forms soluble complex with sodium thiosulfate in photography.

(2) $K[Ag(CN)_2]$ is used for electroplating of silver, and $K[Au(CN)_2]$is used for gold plating.

(3) Some ligands oxidise $Co^{2+}$ to $Co^{3+}$ ion.

(4) EDTA is used for estimation of $Ca^{2+}$ and $Mg^{2+}$in hard water.

(5) Silver and gold are extracted by treating zinc with their cyanide complexes.

## Molecular Recognition and Encapsulation

A molecule is defined as a group of at least two atoms in a definite arrangement held together by covalent chemical bonds. In the narrow use of the word, molecules are electrically neutral. Molecules are distinguished from polyatomic ions in this strict sense. However, in quantum physics, organic chemistry, and biochemistry, the term *molecule* is used less strictly and also is applied to molecular ions, charged organic molecules, and biomolecules. In the kinetic theory of gases, the term *molecule* is often used for any gaseous particle regardless of its composition. According to this definition noble gas atoms are considered molecules despite the fact that they are composed of a single non-bonded atom. A molecule may consist of atoms of a single chemical element, as with oxygen ($O_2$), or of different elements, as with water ($H_2O$). Atoms and complexes connected by non-covalent bonds such as hydrogen bonds or ionic bonds are generally not considered single molecules.

Molecules as components of matter are common in organic substances (and therefore biochemistry). They also make up most of the oceans and atmosphere. A large number of familiar solid substances, however, including most of the minerals that make up the crust, mantle, and core of the Earth itself, contain many chemical bonds, but are *not* made of identifiable molecules. No typical molecule can be defined for ionic crystals (salts) and covalent crystals (network solids), although these are often composed of repeating unit cells that extend either in a plane (such as in graphene) or three-dimensionally (such as in diamond or sodium chloride). The theme of repeated unit-cellular-structure also holds for most condensed phases with metallic bonding. In glasses (solids that exist in a vitreous disordered state), atoms may also be held together by chemical bonds without any definable molecule, but also without any of the regularity of repeating units that characterises crystals.

## Molecular Size

Most molecules are far too small to be seen with the naked eye, but there are exceptions. DNA, a macromolecule, can reach macroscopic sizes, as can molecules of many polymers. The smallest molecule is the diatomic hydrogen ($H_2$), with a length of 0.74 E. Molecules commonly used as building blocks for organic synthesis have a dimension of a few E to several dozen E. Single molecules cannot usually be observed by light (as noted above), but small molecules and even the outlines of individual atoms may be traced in some circumstances by use of an atomic force microscope. Some of the largest molecules are macromolecules or supermolecules.

### *Radius*

*Effective molecular radius* is the size a molecule displays in solution.

### *Molecular Formula*

A compound's empirical formula is the simplest integer ratio of the chemical elements that constitute it. For example, water is always composed of a 2:1 ratio of hydrogen to oxygen atoms, and ethyl alcohol or ethanol is always composed of carbon, hydrogen, and oxygen in a 2:6:1 ratio. However, this does not determine the kind of molecule uniquely – dimethyl ether has the same ratios as ethanol, for instance. Molecules with the same atoms in different arrangements are called isomers. Also carbohydrates, for example, have the same ratio (carbon:hydrogen:oxygen = 1:2:1) (and thus the same empirical formula) but different total numbers of atoms in the molecule. The molecular formula reflects the exact number of atoms that compose the molecule and so characterizes different molecules. However different isomers can have the same atomic composition while being different molecules.

The empirical formula is often the same as the molecular formula but not always. For example the molecule acetylene has molecular formula $C_2H_2$, but the simplest integer ratio of elements is CH.

The molecular mass can be calculated from the chemical formula and is expressed in conventional atomic mass units equal to 1/12th of the mass of a neutral carbon-12 (C isotope) atom. For network solids, the term formula unit is used in stoichiometric calculations.

### *Molecular Geometry*

Molecules have fixed equilibrium geometries—bond lengths and angles— about which they continuously oscillate through vibrational and rotational motions. A pure substance is composed of molecules with the same average geometrical structure. The chemical formula and the structure of a molecule are the two important factors that determine its properties, particularly its reactivity. Isomers share a chemical formula but normally have very different properties because of their different structures. Stereoisomers, a particular type of isomers, may have very similar physico-chemical properties and at the same time different biochemical activities.

### *Molecular Spectroscopy*

Molecular spectroscopy deals with the response (spectrum) of molecules interacting with probing signals of known energy (or frequency, according to Planck's formula). Molecules have quantized energy levels that can be analysed by detecting the molecule's energy exchange through absorbance or emission. Spectroscopy does not generally refer to diffraction studies where particles such as neutrons, electrons, or high energy X-rays interact with a regular arrangement of molecules (as in a crystal).

Spectroscopy was originally the study of the interaction between radiation and matter as a function of wavelength ($\lambda$). Historically, spectroscopy referred to the use of visible light dispersed according to its wavelength, e.g. by a prism. Later the concept was expanded greatly to comprise any measurement of a quantity as a function of either wavelength or frequency. Thus, it also can refer to a response to an alternating field or varying frequency (*í*). A further extension of the scope of the definition added energy (*E*) as a variable, once the very close relationship $E = hí$ for photons was realized (*h* is the Planck constant). A plot of the response as a function of wavelength—or more commonly frequency—is referred to as a spectrum.

Spectrometry is the spectroscopic technique used to assess the concentration or amount of a given chemical (atomic, molecular, or ionic) species. In this case, the instrument that performs such measurements is a spectrometer, spectrophotometer, or spectrograph.

Spectroscopy/spectrometry is often used in physical and analytical chemistry for the identification of substances through the spectrum emitted from or absorbed by them.

Spectroscopy/spectrometry is also heavily used in astronomy and remote sensing. Most large telescopes have spectrographs, which are used either to measure the chemical composition and physical properties of astronomical objects or to measure their velocities from the Doppler shift of their spectral lines.

## Theoretical Aspects

The study of molecules by molecular physics and theoretical chemistry is largely based on quantum mechanics and is essential for the understanding of the chemical bond. The simplest of molecules is the hydrogen molecule-ion, $H_2^+$, and the simplest of all the chemical bonds is the one-electron bond. $H_2^+$ is composed of two positively charged protons and one negatively charged electron, which means that the Schrodinger equation for the system can be solved more easily due to the lack of electron–electron repulsion. With the development of fast digital computers, approximate solutions for more complicated molecules became possible and are one of the main aspects of computational chemistry.

When trying to define rigorously whether an arrangement of atoms is "sufficiently stable" to be considered a molecule, IUPAC suggests that it "must correspond to a depression on the potential energy surface that is deep enough to confine at least one vibrational state". This definition does not depend on the nature of the interaction between the atoms, but only on the strength of the interaction. In fact, it includes weakly bound species that would not traditionally be considered molecules, such as the helium dimer, $He_2$, which has one vibrational bound state and is so loosely bound that it is only likely to be observed at very low temperatures.

The term molecular recognition refers to the specific interaction between two or more molecules through noncovalent bonding such as hydrogen bonding, metal coordination, hydrophobic forces, van der Waals forces, pi-pi interactions, electrostatic and/or electromagnetic effects. The host and guest involved in molecular recognition exhibit molecular complementarity.

### Biological Systems

Molecular recognition plays an important role in biological systems and is observed in between receptor-ligand, antigen-antibody, DNA-protein, sugar-lectin, RNA-ribosome, etc. An important example of molecular recognition is the antibiotic vancomycin that selectively binds with the peptides with terminal D-alanyl-D-alanine in bacterial cells through five hydrogen bonds. The vancomycin is lethal to the bacteria since once it has bound to these particular peptides they are unable to be used to construct the bacteria's cell wall.

### Supramolecular Systems

Chemists have demonstrated that artificial supramolecular systems can be designed that exhibit molecular recognition. One of the earliest examples of such a system are crown ethers which are capable of selectively binding specific cations. However, a number of artificial systems have since been established.

### Static vs. Dynamic

Molecular recognition can be subdivided into *static molecular recognition* and *dynamic molecular recognition*. Static molecular recognition is likened to the interaction between a key and a keyhole; it is a 1:1 type complexation reaction between a host molecule and a guest molecule to form a host-guest complex. To achieve advanced static molecular recognition, it is necessary to make recognition sites that are specific for guest molecules.

In the case of dynamic molecular recognition the binding of the first guest to the first binding site of a host affects the association constant of a second guest with a second binding site. In the case of positive allosteric systems the binding of the first guest increases the association constant of the second guest. While for negative allosteric systems the binding of the first guest decreases the association constant with the second. The dynamic nature of this type of molecular recognition is particularly important since it provides a mechanism to regulate binding in biological systems. Dynamic molecular recognition may enhance the ability to discriminate between several competing targets via the conformational proofreading mechanism. Dynamic molecular recognition is also being studied for application in highly functional chemical sensors and molecular devices.

## Crown Ether

Crown ethers are cyclic chemical compounds that consist of a ring containing several ether groups. The most common crown ethers are oligomers of ethylene oxide, the repeating unit being ethyleneoxy, i.e., -$CH_2CH_2O$-. Important members of this series are the tetramer (n = 4), the pentamer (n = 5), and the hexamer (n = 6). The term "crown" refers to the resemblance between the structure of a crown ether bound to a cation, and a crown sitting on a person's head. The first number in a crown ether's name refers to the number of atoms in the cycle, and the second number refers to the number of those atoms that are oxygen. Crown ethers are much broader than the oligomers of ethylene oxide; an important group are derived from catechol.

Crown ethers strongly bind certain cations, forming complexes. The oxygen atoms are well situated to coordinate with a cation located at the interior of the ring, whereas the exterior of the ring is hydrophobic. The resulting cations often form salts that are soluble in nonpolar solvents, and for this reason crown ethers are useful in phase transfer catalysis. The denticity of the polyether influences the affinity of the crown ether for various cations. For example, 18-crown-6 has high affinity for potassium cation, 15-crown-5 for sodium cation, and 12-crown-4 for lithium cation. The high affinity of 18-crown-6 for potassium ions contributes towards its toxicity.

## Crown Ethers in Nature

Crown ethers are not the only macrocyclic ligands that have affinity for the potassium cation. Ionophores such as valinomycin also display a marked preference for the potassium cation over other cations.

### *History of Synthetic Crown Ethers*

In 1967, Charles Pedersen, who was a chemist working at DuPont, discovered a simple method of synthesizing a crown ether when he was trying to prepare a complexing agent for divalent cations. His strategy entailed linking two catecholate groups through one hydroxyl on each molecule. This linking defines a polydentate ligand that could partially envelop the cation and, by ionization of the phenolic hydroxyls, neutralize the bound dication. He was surprised to isolate a by-product that strongly complexed potassium cation. Citing earlier work on the dissolution of potassium in 16-crown-4, he realized that the cyclic polyethers represented a new class of complexing agents that were capable of binding alkali metal cations. He proceeded to report systematic studies of the synthesis and binding properties of crown ethers in a seminal series of papers. The fields of organic synthesis, phase transfer catalysts, and other emerging disciplines benefited from the discovery of crown ethers. Pedersen particularly popularized the dibenzo crown ethers.

Pedersen shared the 1987 Nobel Prize in Chemistry for the discovery of the synthetic routes to, and binding properties of, crown ethers.

### *Affinity for Cations*

Apart from its high affinity for potassium cations, 18-crown-6 can also bind to protonated amines and form very stable complexes in both solution and the gas phase. Some amino acids, such as lysine, contain a primary amine on their side chains. Those protonated amino groups can bind to the cavity of 18-crown-6 and form stable complexes in the gas phase. Hydrogen-bonds are formed between the three hydrogen atoms of protonated amines and three oxygen atoms of 18-crown-6. These hydrogen-bonds make the complex a stable adduct.

### *Aza-crowns*

21- and 18-membered diazacrown ether derivatives exhibit excellent Calcium and Magnesium selectivities and are widely used in ion-selective electrodes. Some or all of the oxygen atoms in crown ethers can be replaced by nitrogens to form cryptands. A well-known tetrazacrown is cyclen in which there are no oxygens.

## Acyclic Diterpenoids

In chemistry, an open-chain compound (also spelled as open chain compound) is a compound with a linear structure, rather than a cyclic one. An open-chain compound having no side chains is called a straight-chain compound (also spelled as straight chain compound).

***Acetylenes:*** Acyclic (branched or unbranched) and cyclic (with or without side chain) *hydrocarbons* having one or more carbon-carbon triple bonds.

***Acetylides:*** Compounds arising by replacement of one or both hydrogen atoms of acetylene (ethyne) by a metal or other cationic group. E.g. NaCCH monosodium acetylide. NOC Rule C-84.3. By extension, analogous compounds derived from terminal *acetylenes*, RCCH.

***Aliphatic compounds:*** Acyclic or cyclic, saturated or unsaturated carbon compounds, excluding *aromatic compounds.*

***Alkanes:*** Acyclic branched or unbranched *hydrocarbons* having the general formula $C_nH_{2n+2}$, and therefore consisting entirely of hydrogen atoms and saturated carbon atoms. NOC Rule A-1.1. Cf. *cycloalkanes*.

***Alkanium ions:*** *Carbocations* derived from *alkanes* by C-hydronation containing at least one pentacoordinate carbon atom. RNRI Rule RC-82.1.1.2. Cf. *carbonium ions*. E.g. $^+CH_5$ methanium, $[C_2H_7]^+$ ethanium.

***Alkenes:*** Acyclic branched or unbranched *hydrocarbons* having one carbon-carbon double bond and the general formula $C_nH_{2n}$. Acyclic branched or unbranched hydrocarbons having more than one double bond are alkadienes, alkatrienes, etc. NOC Rule A-3.1.

***Alkyl groups:*** Univalent groups derived from *alkanes* by removal of a hydrogen atom from any carbon atom: $C_nH_{2n+1}$-. The groups derived by removal of a hydrogen atom from a terminal carbon atom of unbranched alkanes form a subclass of normal alkyl (*n*-alkyl) groups: $H[CH_2]_n$-. NOC Rule A-1.2. The groups $RCH_2$-, $R_2CH$- (R not equal to H), and $R_3C$- (R not equal to H) are primary, secondary and tertiary alkyl groups respectively.

***Alkyl radicals:*** Carbon-centered radicals derived formally by removal of one hydrogen atom from an *alkane*. RNRI Rule RC-81.1.1. E.g. $CH_3CH_2CH_2\cdot$ propyl.

***Alkylenes:*** 1. An old term, which is not recommended, for *alkenes*, especially those of low molecular weight. 2. An old term for alkanediyl groups commonly but not necessarily having the free valencies on adjacent carbon atoms. E.g. -$CH(CH_3)CH_2$- propylene (systematically called propane-1,2-diyl).

***Alkylidenes:*** *Carbenes* $R_2C$: formed by mono or dialkyl substitution of methylene, $H_2C$: RNRI Rule RC-81.1.3. E.g. $CH_3CH_2CH$: propylidene.

***Alkylidene groups:*** The divalent groups formed from *alkanes* by removal of two hydrogen atoms from the same carbon atom, the free valencies of which are part of a double bond. GNOC Recom. R-2.5. E.g. $(CH_3)_2C$= propan-2-ylidene.

***Alkylidynes:*** *Carbenes* RC formed by alkyl substitution of methyne, HC RNRI Rule RC-81.1.3. E.g. $CH_3CH_2C$ propylidyne.

***Alkynes:*** Acyclic branched or unbranched *hydrocarbons* having a carbon-carbon triple bond and the general formula $C_nH_{2n-2}$, RCCR. Acyclic branched or unbranched hydrocarbons having more than one triple bond are known as alkadiynes, alkatriynes, etc. NOC Rule A-3.2.

***Allenes:*** *Hydrocarbons* (and by extension, derivatives formed by substitution) having two double bonds from one carbon atom to two others: $R_2C{=}C{=}CR_2$. (The simplest member, propadiene, is known as allene.)

***Allylic groups:*** The group $CH_2{=}CHCH_2$- (allyl) and derivatives formed by substitution. The term 'allylic position' or 'allylic site' refers to the saturated carbon atom. A group, such as -OH, attached at an allylic site is sometimes described as 'allylic'.

***Allylic intermediates:*** *Carbanions*, *carbenium ions*, or *radicals*, formally derived by detachment of one hydron, hydride or hydrogen from the $CH_3$ group of propene or derivatives thereof. E.g. $H_2C{=}CHCH_2^+$ allyl cation.

***Cumulenes:*** *Hydrocarbons* (and by extension, derivatives formed by substitution) having three or more cumulative double bonds. E.g. $R_2C{=}C{=}C{=}CR_2$. (Cumulative double bonds are those present in a chain in which at least three contiguous atoms are joined by double bonds.) NOC Rule A-21.1, footnote.

***Dienes:*** Compounds that contain two fixed double bonds (usually assumed to be between carbon atoms). Dienes in which the two double-bond units are linked by one single bond are termed conjugated, e.g. $CH_2{=}CH{-}CH{=}CH_2$ buta-1,3-diene. Dienes in which the double bonds are adjacent are called cumulative, e.g. $CH_3CH{=}C{=}CH_2$ buta-1,2-diene. Those in which one or more of the unsaturated carbon atoms is replaced by a heteroatom may be called heterodienes.

***Hydrocarbons:*** Compounds consisting of carbon and hydrogen only.

***Olefins:*** Acyclic and cyclic *hydrocarbons* having one or more carbon-carbon double bonds, apart from the formal ones in *aromatic compounds*. The class olefins subsumes *alkenes* and cycloalkenes and the corresponding polyenes..

***Paraffins:*** Obsolescent term for saturated *hydrocarbons*, commonly but not necessarily acyclic. Still widely used in the petrochemical industry, where the term designates acyclic saturated hydrocarbons, and stands in contradistinction to *naphthenes*.

***Vinylic groups:*** The vinyl group ($CH_2$=CH-) and derivatives formed by substitution. Informally, a group, such as -OH, attached to the free valence of a (substituted) vinyl group is sometimes referred to as 'vinylic'.

## Terpenoid

The terpenoids, sometimes called isoprenoids, are a large and diverse class of naturally-occurring organic chemicals similar to terpenes, derived from five-carbon isoprene units assembled and modified in thousands of ways. Most are multicyclic structures that differ from one another not only in functional groups but also in their basic carbon skeletons. These lipids can be found in all classes of living things, and are the largest group of natural products.

Plant terpenoids are used extensively for their aromatic qualities. They play a role in traditional herbal remedies and are under investigation for antibacterial, antineoplastic, and other pharmaceutical functions. Terpenoids contribute to the scent of eucalyptus, the flavours of cinnamon, cloves, and ginger, and the colour of yellow flowers. Well-known terpenoids include citral, menthol, camphor, Salvinorin A in the plant *Salvia divinorum*, and the cannabinoids found in *Cannabis*.

The steroids and sterols in animals are biologically produced from terpenoid precursors. Sometimes terpenoids are added to proteins, e.g., to enhance their attachment to the cell membrane; this is known as isoprenylation.

## Structure and Classification

Terpenes are hydrocarbons resulting from the combination of several isoprene units. Terpenoids can be thought of as modified terpenes, wherein methyl groups have been moved or removed, or oxygen atoms added. (Some authors use the term "terpene" more broadly, to include the terpenoids.) Just like terpenes, the terpenoids can be classified according to the number of isoprene units used:

- Hemiterpenoids, 1 isoprene unit (5 Carbons)
- Monoterpenoids, 2 isoprene units (10C)
- Sesquiterpenoids, 3 isoprene units (15C)
- Diterpenoids, 4 isoprene units (20C) (eg ginkgolides)
- Sesterterpenoids, 5 isoprene units (25C)

- Triterpenoids, 6 isoprene units (30C)
- Tetraterpenoids, 8 isoprene units (40C) (eg carotenoids)
- Polyterpenoid with a larger number of isoprene units.

Terpenoids can also be classified according to the number of cyclic structures they contain.

## Biosynthesis

### *Mevalonic Acid Pathway*

Many organisms manufacture terpenoids through the HMG-CoA reductase pathway, the pathway that also produces cholesterol. The reactions take place in the cytosol. The pathway was discovered in the 1950s.

### *MEP/DOXP Pathway*

The 2-*C*-methyl-D-erythritol 4-phosphate/1-deoxy-D-xylulose 5-phosphate pathway (MEP/DOXP pathway), also known as [non-mevalonate pathway] or mevalonic acid-independent pathway, takes place in the plastids of plants and apicomplexan protozoa, as well as in many bacteria. It was discovered in the late 1980s.

Pyruvate and glyceraldehyde 3-phosphate are converted by DOXP synthase (Dxs) to 1-deoxy-D-xylulose 5-phosphate, and by DOXP reductase (Dxr, IspC) to 2-*C*-methyl-D-erythritol 4-phosphate (MEP). The subsequent three reaction steps catalysed by 4-diphosphocytidyl-2-*C*-methyl-D-erythritol synthase (YgbP, IspD), 4-diphosphocytidyl-2-*C*-methyl-D-erythritol kinase (YchB, IspE), and 2-*C*-methyl-D-erythritol 2,4-cyclodiphosphate synthase (YgbB, IspF) mediate the formation of 2-*C*-methyl-D-erythritol 2,4-cyclopyrophosphate (MEcPP). Finally, MEcPP is converted to (*E*)-4-hydroxy-3-methyl-but-2-enyl pyrophosphate (HMB-PP) by HMB-PP synthase (GcpE, IspG), and HMB-PP is converted to isopentenyl pyrophosphate (IPP) and dimethylallyl pyrophosphate (DMAPP) by HMB-PP reductase (LytB, IspH). IPP and DMAPP are the end-products in either pathway, and are the precursors of isoprene, monoterpenoids (10-carbon), diterpenoids (20-carbon), carotenoids (40-carbon), chlorophylls, and plastoquinone-9 (45-carbon). Synthesis of all higher terpenoids proceeds via formation of geranyl pyrophosphate (GPP), farnesyl pyrophosphate (FPP), and geranylgeranyl pyrophosphate (GGPP).

Although both pathways, MVA and MEP, are mutually exclusive in most organisms, interactions between them have been reported in plants and few bacteria species.

# Bibliography


Alexandrov, S.: *Methods for Dissolution, Separation and Preconcentration in Analytical Chemistry,* Narodna Kultura, Sofia, 1995.

Alexandrov, S.; Christova, R.; D. Tsalev, B. Jeliazkova, V. Michaylova: *Practice in Quantitative Analysis*, University Press, Ohridski", 2000.

Arnaudov, M.: *Application of Infrared Spectroscopy in Chemistry,* Nauka i Izkustvo, Sofia, 1981..

Balkwill, Dr. Fran: *DNA Is Here to Stay*, Carolrhoda Books, Inc., Minneapolis, 1993.

Banwell, Colin and Mc. Cash Elaine: *Fundamentals of Molecular Spectroscopy,* Mc Graw-Hill, NY, 1994.

Bard, A.J.; Faulkner, L.R.: *Electrochemical Methods: Fundamentals and Applications*, John Wiley & Sons, New York, 2000.

Bast, R.C. Jr.: *Cancer Medicine*, Decker, Hamilton, 2000.

Bond, A. Curtis: *Modern Polarographic Methods in Analytical Chemistry,* M. Dekker, New York, 1991.

Bontchev, P.; I. Kuleff, M. Mitewa: *Department of Analytical Chemistry 1924–1994 - Bibliography*, University Press, Sofia, 1994.

Bontchev, P.R., *Complexation and Catalytic Activity*, Nauka i Izkustvo, Sofia, 1972.

Bummer, P. M.: *The Science and Practice of Pharmacy*, Mack, Easton, PA, 2000.

Burns, G.; Glazer, A.M.: *Space Groups for Scientists and Engineers,* Academic Press, Boston, 1990.

Callen, H. B.: *Thermodynamics*, Wiley, New York, 1960.

Christian, G. D.: *Analytical Chemistry*, Wiley, NY, 1992.

Christova, R., S. Alexandrov, D. Tsalev, B. Jeljazkova, V. Mihajlova: *Practice in Quantitative Analysis*, Ohridski University Press, Sofia, 2003.

Christova, R.; V. Simeonov: *Potentiometry*, Nauka i Izkustvo, Sofia, 1983.

Crouch, Stanley; Skoog, Douglas A.: *Principles of Instrumental Analysis*, Thomson Brooks/Cole, Australia, 2007.

Cussler, E. L.: *Diffusion: Mass Transfer in Fluid Systems*, Cambridge University Press, Cambridge, UK, 2009.

Dahmen, E. A. M. F.: *Electroanalysis: Theory and Applications in Aqueous and Non-aqueous Media and in Automated Chemical Control*, Elsevier, Amsterdam, 1986.

Dedina, J.; D.L. Tsalev: *Hydride Generation Atomic Absorption Spectrometry*, John Wiley & Sons, Chichester, 1995.

Denbigh, K.: *The Principles of Chemical Equilibrium*, Cambridge Univ. Press, London, 1966.

Dimov, N. K.; Lekova, E. Russeva, I. Havezov, D. Tsalev: *English-Bulgarian Glossary. Instrumental Methods of Analysis*, NIHFI (Pharmaceutical Research Institute), Sofia, 1996.

Duarte, F. J. and L.W. Hillman: *Dye Laser Principles,* Academic, New York, 1990.

Duxbury, G.: *Infrared Vibration-Rotation Spectroscopy, From Free Radicals to the Infrared Sky*, Wiley, NJ, 1999.

Evans, A.: *Potentiometry and Ion Selective Electrodes*, Wiley, New York, 1987.

Gardiner, W. P.: *Statistical Analysis Methods for Chemists: A software-based approach*, Royal Society of Chemistry, Cambridge, UK, 1997.

Guyton, A.C.; Hall, J.E.: *Textbook of Medical Physiology*, Saunders, Philadelphia, 2006.

Hala, E., I. Wichterle, J. Polak, and T. Boublik: *Vapor–Liquid Equilibrium at Normal Pressures,* Pergamon, Oxford, 1968.

Havezov, I.P.; D.L. Tsalev: *Atomic Absorption Analysis*, Nauka i Izkustvo, Sofia, 1980.

Holland, C. D.: *Fundamentals of Multicomponent Distillation,* McGraw-Hill, New York, 1981.

Holler, F. James; Skoog, Douglas A., West, Donald M.: *Fundamentals of Analytical Chemistry*, Saunders College Pub., Philadelphia, 1996.

Horsely, L. H.: *Azeotropic Data, ACS Advances in Chemistry*, American Chemical Society, Washington, DC, 1952.

Hubert H. Girault: *Analytical and Physical Electrochemistry*, EPFL. , Lausanne, 2004.

Humphrey, J. L,. and G. E. Keller II: *Separation Process Technology*, McGraw-Hill, New York, 1997.

Jeliazkova, B.; Ch. Daiev, S. Beltchev, D. Tsalev, L. Shishkova, M. Mitewa, Yu. Charizanov, N. Piperov: *Practice in Physical Methods of Analysis and Investigation of Inorganic Materials*, University Press, Ohridski, 1991.

Jones, Denny: *Principles and Prevention of Corrosion,* Upper Saddle River, Prentice Hall, New Jersey, 1996.

Jornitz, Maik W.: *Sterile Filtration*, Springer, Germany, 2006.

Joseph Needham: *Science & Civilisation in China,* Cambridge University Press, Cambridge, England, 1962.

Joseph, C. Salamone: *Polymeric Materials Encyclopedia,* CRC Press, Boca Raton, FL, 1996.

Karger, B. L., L. R. Snyder, and C. Horvath: *An Introduction to Separation Science*, John Wiley and Sons, New York, 1973.

Kenneth I. Ozomwna: *Recent Advances in Analytical Electrochemistry 2007*, Transworld Research Network, 2007.

Kittel and Kroemer: *Thermal Physics*, W. H. Freeman, New York, 1980.

Kotcheva, L.L.: D. Chr. Nonova: *Qualitative Semi-Micro Analysis*, University Press, Ohridski, Sofia, 2000.

Kuleff, I.; R. Djingova: *75th Anniversary of the Department of Analytical Chemistry 1924–1994 - Bibliography 1994–1999* , University Press, Sofia, 1999.

Kuleff, I.; R. Zlatanova, P. Konstantinov, M. Dimitrov, G.Genchev , P. Peev: *How Dangerous for the Population Are the Liquid Radioactive Releases from NPP "Kozloduy"?*, BalBok, Sofia, 1994..

Laidler, Keith: *The World of Physical Chemistry*, Oxford University Press, Oxford, 1993.

Lindon, John C.: *Encyclopedia of Spectroscopy and Specrometry*, Academic Press, Uk, 1999.

Massart, D. L.: *Chemometrics: A Textbook*, Amsterdam, Elsevier, 1988.

Miller, J. N. and J. C. Miller: *Statistics and Chemometrics for Analytical Chemistry*, Prentice Hall, UK, 2000.

Nicholas P.: *Hazardous Chemicals in the Polymer Industry,* M. Dekker, Cheremisinoff, New York, 1995.

Nieman, Timothy A.; Skoog, Douglas A.; Holler, F. James: *Principles of Instrumental Analysis*, Brooks/Cole, Pacific Grove, CA, 1998.

Nikolov, G.; P.R. Bontchev: *Mechanisms of Inorganic Reactions in Solutions*, Nauka i Izkustvo, Sofia, 1969.

Norman, A. Waterman and Michael F. Ashby.: *The Materials Selector,* CRC Press, Boca Raton, 1997.

Odian, George: *Principles of Polymerization,* John Wiley & Sons, New York, 1991.

Olagoke Olabisi: *Handbook of Thermoplastics,* Marcel Dekker, New York, 1997.

Perry, R. H. and D. Green: *Perry's Chemical Engineer's Handbook*, McGraw-Hill, New York, 1984.

Prausnitz, J. M., R. N. Lichtenthaler, and E. G. de Azevedo: *Molecular Thermodynamics of Fluid-Phase Equilibria*, Prentice-Hall, Englewood Cliffs, NJ, 1986.

Prichard, E.: *Analytical Measurement Terminology*, LGC/Royal Society of Chemistry, Cambridge, UK, 2001.

Reid, R. C., J. M. Prausnitz, and B. E. Poling: *The Properties of Gases and Liquids*, McGraw-Hill, New York, 1987.

Seeman, N.C.: *Biomolecular Stereodynamics*, Adenine Press, New York, 1981.

Simeonov, V.: *Information Aspects in Chemical Analysis*, Obrazovatelni Tekhnologii, Sofia, 1999.

Skoog, D.A.: *Principles of Instrumental Analysis*, Saunders College Publishing, 1991.

Spasov, St.; M. Arnaudov: *Application of Spectroscopy in Organic Chemistry*, Nauka i Izkustvo, Sofia, 1978.

Stone, D. C.: *Instrumental Analysis Laboratory Manual*, Department of Chemistry, University of Toronto, 2005.

Stuart, M. Lee: *International Encyclopedia of Composites,* VCH Publishers, New York, 1990.

Tsalev, D.L.: *Atomic Absorption Spectrometry in Occupational and Environmental Health Practice*, Vol. II: *Determination of Individual Elements*, CRC Press, Boca Raton, Florida, 1984.

Tsalev, D.L.; I.P. Havezov: *Atomic Absorption Analysis*, Khimia, Leningrad, 1983.

Walas, S. M.: *Phase Equilibria in Chemical Engineering,* Butterworths, Reading, MA, 1985.

Wang, Joseph C.: *Analytical Electrochemistry*, John Wiley & Sons, Chichester, 2000.

Workman, Jerome; Workman, Jerry: *Handbook of Organic Compounds: NIR, IR, Raman, and UV-Vis Spectra Featuring Polymers and Surfactants*, Academic Press, San Diego, 2001.

Zaprianov, Z.K.: *Atomic Absorption Spectrometry in Occupational and Environmental Health Practice*, Vol. I: *Analytical Aspects and Health Significance*, CRC Press, Boca Raton, Florida, 1983.

# Index